Springer Series in Reliability Engineering

Series Editor

Hoang Pham, Department of Industrial and Systems Engineering, Rutgers University, Piscataway, NJ, USA

More information about this series at http://www.springer.com/series/6917

Chao Chen · Genserik Reniers · Ming Yang

Integrating Safety and Security Management to Protect Chemical Industrial Areas from Domino Effects

Springer

Chao Chen
Safety and Security Science Group
Delft University of Technology
Delft, The Netherlands

Genserik Reniers
Safety and Security Science Group
Delft University of Technology
Delft, The Netherlands

Ming Yang
Safety and Security Science Group
Delft University of Technology
Delft, The Netherlands

ISSN 1614-7839 ISSN 2196-999X (electronic)
Springer Series in Reliability Engineering
ISBN 978-3-030-88910-4 ISBN 978-3-030-88911-1 (eBook)
https://doi.org/10.1007/978-3-030-88911-1

This Springer imprint is published by the registered company Springer Nature Switzerland AG
The registered company address is: Gewerbestrasse 11, 6330 Cham, Switzerland

Preface

In recent decades, domino effect accidents in the process industry have raised an increasing concern in scientific, technical, and management domains. Domino effects such as the Buncefield accident in 2005 are characterized by a chain of accidents and may result in extremely severe consequences. The primary event can be an accident, a natural disaster, or an intentional attack. To prevent and mitigate domino effects, growing research on risk assessment and management of domino effects has been conducted in recent decades such as quantitative risk assessment of domino effects and safety barrier management for preventing domino effects. However, little attention has been paid to domino effects triggered by intentional events (intentional domino effects) such as terrorist attacks. This book thus aims to introduce advanced approaches that can model and manage both intentional and unintentional domino effects, developing a safer, securer, and more resilient chemical plant.

Chapter 1 provides a systematic literature review of domino effects in the process industry by classifying the methods used for modeling and managing domino effects, analyzing current research trends, and discussing future research needs. The models are divided into three categories (analytical methods, graphical methods, and simulation methods) and the management strategies are grouped into five types (inherent safety, management of safety barriers, emergency response, cooperative prevention, and security strategies). These approaches and strategies are analyzed and discussed to identify the research gaps and explore future research directions.

Chapter 2 introduces a dynamic graph approach for modeling the spatial-temporal evolution of fire-induced domino effects. The core of this section is the developed Domino Evolution Graph (DEG) model and the Minimum Evolution Time (MET) algorithm for solving the model. A case study is provided to test the model while another case is used to show its application in chemical clusters with a large number of hazardous installations. Compared with previous probabilistic models, the DEG model concerns more physical mechanisms and it is more flexible and visible to model the dynamic escalation process.

Chapter 3 provides a dynamic evolution model of VCE induced domino effects. Past risk assessment methods on VCEs ignore the effects of vapor cloud dispersion and delayed ignitions on the vulnerability of installations. As a result, a Dynamic VCE

Evolution Assessment (DVEA) model is developed based on a dynamic event tree, considering the spatial-temporal evolution of VCEs and the uncertainty of Delayed Ignition Time (DIT). Multiple ignition sources can be considered in this model, addressing the uncertainties of ignition time and ignition positions.

Chapter 4 introduces a dynamic approach, "Dynamic Graph Monte Carlo" (DGMC), for modeling multiple accident scenarios in domino effects. In this chapter, a chemical plant is modeled as a multi-agent system (installations, humans, and ignition sources), and the vulnerabilities of humans exposed to toxic gas, fire, and VCE are considered. Since the spatial-temporal evolution of multiple accident scenarios is addressed, the DGMC model can avoid underestimating domino effect risk, the damage risk of installations and humans.

Chapter 5 introduces an integrated safety and security management framework to tackle both intentional and unintentional domino effects. This framework mainly consists of six steps: chemical plant description, threat and hazard identification, the vulnerability of installations subject to hazards and threats, the vulnerability of installations exposed to domino effects, consequence analysis, risk treatment, and risk reduction. According to this framework, protection strategies encompassing both safety and security can be formulated to reduce domino effect risk.

Chapter 6 provides a cost-benefit management approach for optimizing the investment and allocation of safety and security resources based on the integrated safety and security management framework. A cost-benefit analysis is used to quantify the performance of protection strategies on the prevention and mitigation of domino effects. An optimization algorithm called "PROTOPT" is developed to achieve the most "profitable" protection strategy.

Chapter 7 introduces a resilience concept in domino effect management. A stochastic dynamic method is established to quantify the resistant capability, the mitigation capability, the adaption capability, and the restoration capability of chemical plants, supporting the allocation of safety barriers, security barriers, adaption measures, and restoration measures. Once a domino effect is inevitable, a resilient chemical plant may rapidly restore from the escalation disaster and reduce the losses.

Chapter 8 concludes this book and discusses future research on domino effect management in the process industry.

Delft, The Netherlands

Chao Chen
Genserik Reniers
Ming Yang

Contents

List of Figures

List of Tables

Chapter 1
Safety and Security of Domino Effects in the Process Industry: The State of the Art

1.1 Introduction

In the chemical process industry (CPI),[1] raw materials are processed into finished products by chemical and physical conversions while many hazardous materials are being stored, transferred, and processed in different facilities [3]. These facilities situated nearby one another pose a risk to each other due to possible major accident scenarios caused by the release of hazardous materials in any of the facilities. Once a major accident scenario occurs, it may sequentially or simultaneously propagate to nearby installations, leading to domino effects and possibly resulting in catastrophic consequences. Domino effects may be considered high-impact, low-frequency (HILF) events [4], a type II event [5], and they are responsible for several catastrophic disasters in the chemical industry [1, 6]. For instance, one of the most well-known domino effect accidents took place in Mexico City, in 1984, killing more than 500 people and injuring 7000 people [7]. Most recently, on March 21, 2019, an escalation accident occurred at Jiangsu Tianjiayi Chemical Company, China, resulting in at least 78 deaths, 617 injuries, and massive property loss [8].

In light of the severe consequences caused by domino effects, increasing attention has been paid to the prevention and mitigation of escalation effects since the 1990s [9, 10]. The early research work focused on the risk assessment and management of domino effects induced by accidental events [11–13]. In 2008, Reniers et al. [14] developed a methodology to manage domino effects caused by intentional attacks. Since then, intentional domino effects have also attracted attention in the scientific and technical domain [15, 16]. Since some years now, academic publications on domino effects triggered by natural disasters are available in the literature [17, 18]. These studies draw public concerns on preventing domino effects and promote the inclusion of domino effects into safety laws, regulations, and recommendations. In

[1] This chapter is mainly based on two publications: Chen et al. [1], Chen [2].

C. Chen et al., *Integrating Safety and Security Management to Protect Chemical Industrial Areas from Domino Effects*, Springer Series in Reliability Engineering,
https://doi.org/10.1007/978-3-030-88911-1_1

1996, the EU issued the "Seveso-II" Directive (Directive 96/82/EC) in which chemical companies are required to assess "domino" accident hazards inside and outside chemical industrial areas [19]. In 2012, the "Seveso-III" Directive emphasized the importance of exchanging information between chemical plants to prevent external domino effects in chemical clusters [20].

Besides, literature reviews related to past accident statistics [21–23], risk assessment of domino effects [4], escalation thresholds [24], bibliometric analysis [25], and historical summaries [26] are also available in the literature. However, intentional domino effects and escalation effects caused by natural disasters have almost been ignored in these reviews. Therefore, this chapter is to conduct a systematic review on domino effect assessment and management, obtaining insight into modeling approaches and protection strategies. Besides, the evolution of these models and methods are analyzed, and the main concerns for future needs are also discussed.

1.2 Method and Materials

This literature review focuses on the research issues and approaches related to risk assessment and management of domino effects in the chemical process industry. According to this study, we expect to achieve the following research objectives:

(1) A classification of current research literature
(2) A summary of the current classifications of domino effects
(3) The current models for risk assessment of domino effects
(4) The current management strategies and approaches used for decision-making on preventing and mitigating domino effects
(5) The research gaps related to domino effect assessment and management.

To obtain the above research objectives, a four-step literature review method based on a systematic review and meta-synthesis techniques [27] is proposed.

Figure 1.1 shows the procedures of the proposed method. The review method consists of six steps. The first step is to search extensive literature from the online library of the Delft University of Technology. The literature is searched and collected from two databases: Core Collection of Web of Science (WoS) and ScienceDirect. The searching topics to serve as input in search engines are as follows:

Topic: domino effect OR knock-on event OR catastrophic effect, OR chain of accidents OR escalating event;

AND Topic: process industry OR chemical industry OR chemical plant OR chemical industrial cluster OR chemical industrial park OR oil OR gas OR petroleum OR LNG OR "LPG".

The literature collection date was on April 29th, 2019. A total of 284 publications were obtained. The title and abstract of each publication were examined thoroughly to further refine the publications, excluding the publications that were not closely related to domino effects. Finally, 127 journal papers were selected as the literature

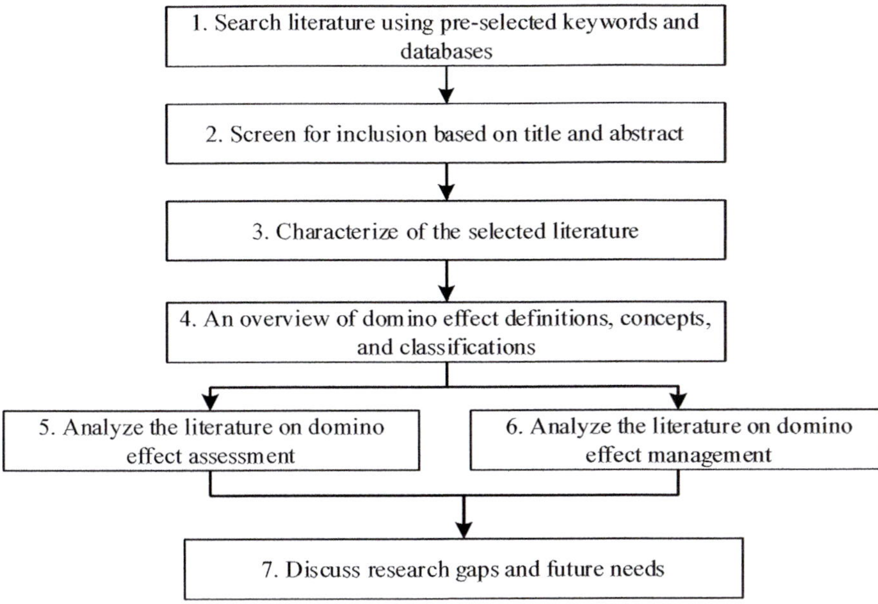

Fig. 1.1 Procedures of literature review

for analysis. These articles were from 32 international journals such as Journal of Loss Prevention in the Process Industries, Journal of Hazardous Materials, Process Safety and Environmental Protection, Reliability Engineering & System Safety, and Safety Science. Among these papers, 57 of them were published in the past five years (from 2015 to 2019) indicating that domino effects have obtained increasing attention in the scientific domain. The most productive authors in this domain include Valerio Cozzani, Genserik Reniers, Nima Khakzad, Gabriele Landucci, and Faisal Khan.

1.3 Characterization of Selected Publications

To analyze the research related to domino effects, the selected 139 journal papers are characterized by their research topics, research issues, and research approaches, as shown in Fig. 1.2.

The research topics are divided into two categories: domino effect assessment and domino effect management. The former consists of 2 research issues (e.g., vulnerability assessment and risk assessment), while the latter includes five research issues (e.g., inherent safety, safety barrier management, emergency management, cooperative management, security management). Research approaches used in each research issue are also identified. Based on the characterization, each paper can be identified

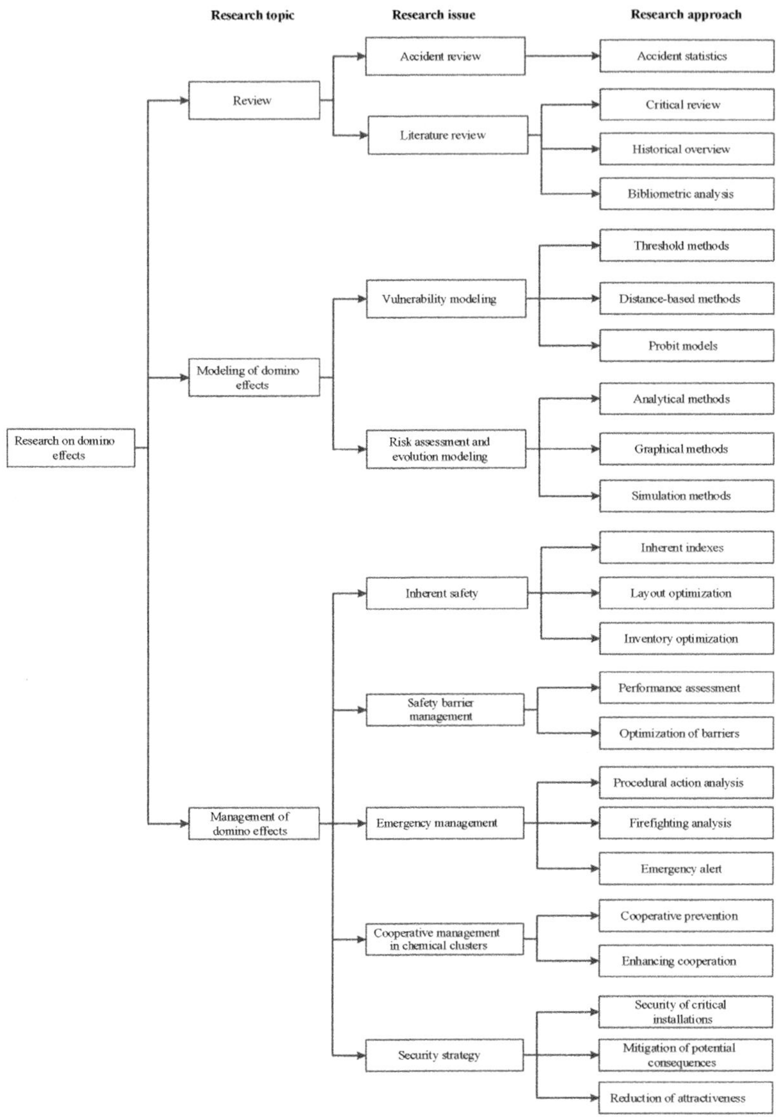

Fig. 1.2 Characterization of domino effect research (Chen et al. [1])

by its research topics (2), research issues (7), and research approaches (19). As a result, the selected 139 journal papers are divided into 23 categories, as shown in Table 1.1. The rest of this chapter will analyze and discuss research papers according to this classification.

Table 1.1 Characterization of current researches (Chen et al. [1])

Publication	Topic	Research issue	Research approach
Darbra et al. (2010)	Review	Accident	Accident statistics
Abdolhamidzadeh et al. (2011)	Review	Accident	Accident statistics
Hemmatian et al. (2014)	Review	Accident	Accident statistics
Alileche et al. (2015)	Review	Literature	Critical review
Necci et al. (2015)	Review	Literature	Critical review
Swuste et al. (2019)	Review	Literature	Historical overview
Li et al. (2017)	Review	Literature	bibliometric analysis
Alileche et al. (2015)	Modeling	Vulnerability	Threshold methods
Cozzani et al. (2006b)	Modeling	Vulnerability	Threshold methods
Salzano and Cozzani (2006)	Modeling	Vulnerability	Threshold methods
Cozzani and Salzano (2004c)	Modeling	Vulnerability	Probabilistic methods
Gubinelli et al. (2004)	Modeling	Vulnerability	Probabilistic methods
Gubinelli and Cozzani (2009b)	Modeling	Vulnerability	Probabilistic methods
Gubinelli and Cozzani (2009a)	Modeling	Vulnerability	Probabilistic methods
Hauptmanns (2001a)	Modeling	Vulnerability	Probabilistic methods
Hauptmanns (2001b)	Modeling	Vulnerability	Probabilistic methods
Jia et al. (2017)	Modeling	Vulnerability	Probabilistic methods
Jujuly et al. (2015)	Modeling	Vulnerability	Probabilistic methods
Landucci et al. (2009a)	Modeling	Vulnerability	Probabilistic methods
Landucci et al. (2015b)	Modeling	Vulnerability	Probabilistic methods
Lisi et al. (2014)	Modeling	Vulnerability	Probabilistic methods
Lisi et al. (2015)	Modeling	Vulnerability	Probabilistic methods
Mukhim et al. (2017)	Modeling	Vulnerability	Probabilistic methods
Pula et al. (2007)	Modeling	Vulnerability	Probabilistic methods
Salzano and Cozzani (2005)	Modeling	Vulnerability	Probabilistic methods
Salzano et al. (2014)	Modeling	Vulnerability	Probabilistic methods
Sun et al. (2012)	Modeling	Vulnerability	Probabilistic methods
Sun et al. (2013b)	Modeling	Vulnerability	Probabilistic methods
Sun et al. (2013a)	Modeling	Vulnerability	Probabilistic methods
Sun et al. (2016b)	Modeling	Vulnerability	Probabilistic methods
Sun et al. (2017)	Modeling	Vulnerability	Probabilistic methods
Tugnoli et al. (2014b)	Modeling	Vulnerability	Probabilistic methods
Tugnoli et al. (2014a)	Modeling	Vulnerability	Probabilistic methods
Zhang and Jiang (2008)	Modeling	Vulnerability	Probabilistic methods

(continued)

Table 1.1 (continued)

Publication	Topic	Research issue	Research approach
Zhang and Chen (2009)	Modeling	Vulnerability	Probabilistic methods
Ahmadi et al. (2019)	Modeling	Vulnerability	CFD/FEM methods
Argentia et al. (2014)	Modeling	Vulnerability	CFD/FEM methods
Landucci et al. (2009b)	Modeling	Vulnerability	CFD/FEM methods
Landucci et al. (2016b)	Modeling	Vulnerability	CFD/FEM methods
Rum et al. (2018)	Modeling	Vulnerability	CFD/FEM methods
Antonioni et al. (2009)	Modeling	Risk assessment	Analytical method
Antonioni et al. (2015)	Modeling	Risk assessment	Analytical method
Baesi et al. (2013)	Modeling	Risk assessment	Analytical method
Bagster and Pitblado (1991)	Modeling	Risk assessment	Analytical method
Cozzani et al. (2006a)	Modeling	Risk assessment	Analytical method
Cozzani et al. (2005)	Modeling	Risk assessment	Analytical method
Cozzani and Salzano (2004b)	Modeling	Risk assessment	Analytical method
Cozzani and Salzano (2004a)	Modeling	Risk assessment	Analytical method
Cozzani et al. (2014)	Modeling	Risk assessment	Analytical method
Kadri et al. (2013)	Modeling	Risk assessment	Analytical method
Khan and Abbasi (1998a)	Modeling	Risk assessment	Analytical method
Khan and Abbasi (2001)	Modeling	Risk assessment	Analytical method
Khan and Abbasi (1996)	Modeling	Risk assessment	Analytical method
Khan and Abbasi (1998b)	Modeling	Risk assessment	Analytical method
Khan and Abbasi (2000)	Modeling	Risk assessment	Analytical method
Khan et al. (2001a)	Modeling	Risk assessment	Analytical method
Khan et al. (2001b)	Modeling	Risk assessment	Analytical method
Ramirez-Camacho et al. (2015)	Modeling	Risk assessment	Analytical method
Silva et al. (2016)	Modeling	Risk assessment	Analytical method
van der Voort et al. (2007)	Modeling	Risk assessment	Analytical method
Zhang and Chen (2013)	Modeling	Risk assessment	Analytical method
Zhou and Reniers (2018a)	Modeling	Risk assessment	Analytical method
Alileche et al. (2017)	Modeling	Risk assessment	Graphical method
Chen et al. (2018)	Modeling	Risk assessment	Graphical method
Dai et al. (2019)	Modeling	Risk assessment	Graphical method
Ji et al. (2018)	Modeling	Risk assessment	Graphical method
Kamil et al. (2019)	Modeling	Risk assessment	Graphical method
Khakzad (2015)	Modeling	Risk assessment	Graphical method
Khakzad (2018b)	Modeling	Risk assessment	Graphical method

(continued)

Table 1.1 (continued)

Publication	Topic	Research issue	Research approach
Khakzad (2019)	Modeling	Risk assessment	Graphical method
Khakzad et al. (2018a)	Modeling	Risk assessment	Graphical method
Khakzad et al. (2018b)	Modeling	Risk assessment	Graphical method
Khakzad et al. (2013)	Modeling	Risk assessment	Graphical method
Khakzad and Reniers (2015b)	Modeling	Risk assessment	Graphical method
Khakzad et al. (2016)	Modeling	Risk assessment	Graphical method
Reniers and Dullaert (2007)	Modeling	Risk assessment	Graphical method
Yuan et al. (2016)	Modeling	Risk assessment	Graphical method
Yang et al. (2018)	Modeling	Risk assessment	Graphical method
Zhou and Reniers (2017b)	Modeling	Risk assessment	Graphical method
Abdolhamidzadeh et al. (2010)	Modeling	Risk assessment	Simulation method
Ahmed et al. (2012)	Modeling	Risk assessment	Simulation method
Rad et al. (2014)	Modeling	Risk assessment	Simulation method
Zhang et al. (2018)	Modeling	Risk assessment	Simulation method
Cozzani et al. (2007)	Management	Inherent safety	Inherent safety indexes
Cozzani et al. (2009)	Management	Inherent safety	Inherent safety indexes
Landucci et al. (2008)	Management	Inherent safety	Inherent safety indexes
Tugnoli et al. (2008b)	Management	Inherent safety	Inherent safety indexes
Tugnoli et al. (2008a)	Management	Inherent safety	Inherent safety indexes
Bernechea and Arnaldos (2014)	Management	Inherent safety	Layout optimization
Dan et al. (2015)	Management	Inherent safety	Layout optimization
de Lira-Flores et al. (2014)	Management	Inherent safety	Layout optimization
de Lira-Flores et al. (2018)	Management	Inherent safety	Layout optimization
Jung et al. (2011)	Management	Inherent safety	Layout optimization
Khakzad and Reniers (2015a)	Management	Inherent safety	Layout optimization
Latifi et al. (2017)	Management	Inherent safety	Layout optimization
Lee et al. (2005)	Management	Inherent safety	Layout optimization
Lee et al. (2006)	Management	Inherent safety	Layout optimization
López-Molina et al. (2013)	Management	Inherent safety	Layout optimization
Nomen et al. (2014)	Management	Inherent safety	Layout optimization
So et al. (2011)	Management	Inherent safety	Layout optimization
Khakzad et al. (2014)	Management	Inherent safety	Inventory optimization

(continued)

Table 1.1 (continued)

Publication	Topic	Research issue	Research approach
Bucelli et al. (2018)	Management	Safety barriers	Performance assessment
Janssens et al. (2015)	Management	Safety barriers	Performance assessment
Khakzad et al. (2017a)	Management	Safety barriers	Performance assessment
Khakzad et al. (2017c)	Management	Safety barriers	Performance assessment
Landucci et al. (2015a)	Management	Safety barriers	Performance assessment
Landucci et al. (2016a)	Management	Safety barriers	Performance assessment
Landucci et al. (2017a)	Management	Safety barriers	Performance assessment
Landucci et al. (2017b)	Management	Safety barriers	Performance assessment
Sun et al. (2016a)	Management	Safety barriers	Performance assessment
Tugnoli et al. (2012)	Management	Safety barriers	Performance assessment
Tugnoli et al. (2013)	Management	Safety barriers	Performance assessment
Ghasemi and Nourai (2017)	Management	Safety barriers	Optimization of barriers
Khakzad and Reniers (2017)	Management	Safety barriers	Optimization of barriers
Khakzad et al. (2017b)	Management	Safety barriers	Optimization of barriers
Khakzad et al. (2018c)	Management	Safety barriers	Optimization of barriers
Tsai et al. (2018)	Management	Emergency	Procedural action analysis
Zhou et al. (2016)	Management	Emergency	Procedural action analysis
Zhou and Reniers (2016)	Management	Emergency	Procedural action analysis
Zhou and Reniers (2017a)	Management	Emergency	Procedural action analysis
Zhou and Reniers (2018b)	Management	Emergency	Procedural action analysis
Cincotta et al. (2019)	Management	Emergency	Firefighting analysis

(continued)

Table 1.1 (continued)

Publication	Topic	Research issue	Research approach
Khakzad (2018a)	Management	Emergency	Firefighting analysis
Khakzad (2018d)	Management	Emergency	Firefighting analysis
Hosseinnia et al. (2018)	Management	Emergency	Emergency alert
Reniers et al. (2005a)	Management	Cooperative	Cooperative prevention
Reniers et al. (2005b)	Management	Cooperative	Cooperative prevention
Reniers et al. (2009)	Management	Cooperative	Cooperative prevention
Reniers and Soudan (2010)	Management	Cooperative	Cooperative prevention
Pavlova and Reniers (2011)	Management	Cooperative	Enhancing cooperation
Reniers (2010)	Management	Cooperative	Enhancing cooperation
Reniers et al. (2012)	Management	Cooperative	Enhancing cooperation
Reniers et al. (2008)	Management	Security	Security of critical installations
Reniers et al. (2014)	Management	Security	Security of critical installations
Khakzad and Reniers (2019)	Management	Security	Mitigation of consequences
Reniers and Audenaert (2014)	Management	Security	Mitigation of consequences
Srivastava and Gupta (2010)	Management	Security	Mitigation of consequences
Chen et al. (2019)	Management	Security	Reduction of attractiveness
Khakzad (2018c)	Management	Security	Reduction of attractiveness

1.4 An Overview of Domino Effect Definitions, Characteristics and Classifications

Based on the review papers on domino effects, this section illustrates domino effect definitions, characteristics, and classifications.

1.4.1 Domino Effect Definitions

In the Collins dictionary, the item "domino effect" refers to the phenomenon that one event causes another similar event, which in turn causes another event, and so on. In the broad definition, the event can be desired or undesired while the event is specially defined as an undesired event in the safety and security domain. In terms of process safety, many definitions of "domino effects" are provided in the review literature [1, 4, 26]. Table 1.2 lists several definitions of "domino effects".

Each definition provided in Table 1.2 summarizes one or more characteristics of domino effects. For example, the definition provided by Lees [28] describes the spatial propagation of domino effects (from one unit to another unit) while Khan and Abbasi [30] highlights that the primary accident may cause accidents in multiple units. Due to the propagation, the consequences of domino effects are more severe than those of the primary event, which is characterized by the definitions provided by Council Directive 96/82/EC [29], Reniers et al. [31], Darbra et al. [21], etc. Reniers and Cozzani [6] developed a more comprehensive definition, covering five main characteristics of domino effects:

(i) A "primary event," initiating the domino effect;
(ii) Escalation vectors responsible for possible accident propagation;
(iii) One or more secondary accident events;

Table 1.2 A list of the definitions of "domino effects"

Sources	Definitions
Lees [28]	An event in one unit that causes a follow-up event in another unit
Council Directive 96/82/EC [29]	Stablishments are sited in such a way or so close together as to increase the probability and possibility of major accidents or aggravate their consequences
Khan and Abbasi [30]	An accident in one unit causes accidents in one or more other units
Reniers et al. [31]	A cascade of events in which the consequences of a previous accident are increased by following one(s), spatially as well as temporally, leading to a major accident
Khan and Abbasi [30]	A chain of accidents that a fire/explosion/missile/toxic load generated by an accident in one unit in an industry causes secondary and higher-order accidents in other units
Darbra et al. [21]	A relatively minor accident can initiate a sequence of events that cause damage over a much larger area and lead to far more severe consequences
Abdolhamidzadeh et al. [23]	A loss of containment accident in a process unit becomes the trigger of one or more loss of containment accidents in one or more other process units
Reniers and Cozzani [6]	A primary unintentional or intentional event propagates within an equipment ('temporally'), or/and to nearby equipment ('spatially'), sequentially or simultaneously, triggering one or more secondary unwanted events, in turn possibly triggering (higher-order) unwanted events, resulting in overall consequences more severe than those of the primary event

(iv) The overall consequences are more severe than those of the primary event;
(v) The primary event can be intentional or unintentional.

Since this work concerns both intentional domino effects and unintentional domino effects, the comprehensive definition provided by Reniers and Cozzani [6] is adopted in the book.

1.4.2 Domino Effect Characteristics

Domino effect characteristics can be analyzed and summarized from past domino effect accidents [22, 23]. Abdolhamidzadeh et al. [23] analyzed 224 domino effect accidents between 1910 to 2008 in the process industry. They demonstrated that 43% of the recorded domino accidents were induced by fires, and 53% of those were triggered by explosions. Although the most common major accident scenarios in the process industry include fire, explosion, and toxic cloud dispersion, the last scenario is always neglected since they do not directly lead to damage to secondary installations [32]. Among the domino events triggered by fires, pool fire (80%) was the most frequent starting scenario [23]. Among explosions, VCE (vapor cloud explosion) was the most frequent scenario. Besides, long-lasting stationary fires (i.e., pool fires and jet fires) are responsible for most of the escalation events in industrial accidents [33]. The analysis further indicated that 44% of jet fire accidents occurred in transportation, 36% in process plants, 11% during loading/unloading operations, and 9% in storage plants. According to the analysis, primary events of domino effects can be divided into three categories: fires, explosions. The physical effects induced by primary events are called escalation vectors responsible for possible hazardous scenario propagations [11]. The escalation vectors induced explosions mainly consist of overpressure and fragment projection, while these induced by fires include heat radiation, fire, and impingement. Possible primary scenarios and escalation vectors derived from past domino accident analyses [10, 34] are shown in Table 1.3.

Table 1.3 Possible escalation vectors of different primary scenarios (Chen et al. [1])

Primary scenario	Escalation vector
Pool fire	Heat radiation, fire impingement
Jet fire	Heat radiation, fire impingement
Fireball	Heat radiation, fire impingement
Flash fire	Fire impingement
BLEVE	Overpressure, fragment projection
Confined explosion	Overpressure, fragment projection
Mechanical explosion	Overpressure, fragment projection
VCE	Overpressure

BLEVE: Boing Liquid Expanding Vapor Explosion; VCE: Vapor Cloud Explosion

Hemmatian et al. [22] conducted a historical survey on 330 accidents related to domino effects in process/storage plants and hazardous material transportations. This study indicates that process plants (39%) and storage areas (33%) are the most common locations for domino effects. The causes of domino effects involve mechanical failure, external events, human factors, etc. The distribution of domino effect causes is shown in Fig. 1.3 [22].

According to Fig. 1.3, the most frequent cause is mechanical failure (35.2%), followed by external events (29.4%), human error (24.6%), and impact failure (16.7). External events include Natech events (6.1%) and sabotage events (0.6%). The Natech events include 14 lightning events, 3 extreme temperature events, 2 earthquakes, and 2 floodings. Sabotage events are intentional, while other causes are unintentional. As a result, domino effects can be divided into two categories: unintentional domino effects (the causes include mechanical failure, human error, aging, natural disasters, etc.) and intentional domino effects (the causes involve terrorist attacks, sabotage, criminal actions, etc.). A comparison between intentional domino effects and unintentional domino effects is provided in Table 1.4.

In the process industry, major accidents may be triggered by natural disasters, resulting in the damage of installations and the loss of containment (LOC) of hazardous substances, which are called Natechs [35]. If we consider the LOC as a secondary event, Natechs may be regarded as a special domino effect triggered by natural events such as lightning, earthquakes, floods, and hurricanes. In this book, domino effects triggered by Natech events are narrowly tailored as Natech events in which the damaged equipment furtherly causes escalation and results in major accident scenarios at other hazardous installations [18].

The frequencies of domino effects caused by intentional attacks are much less than those triggered by accidental events and natural disasters, but the overall consequences may be more severe due to simultaneous damage of installations induced by multiple-target attacks [15, 16]. Besides, safety barriers may also be attacked,

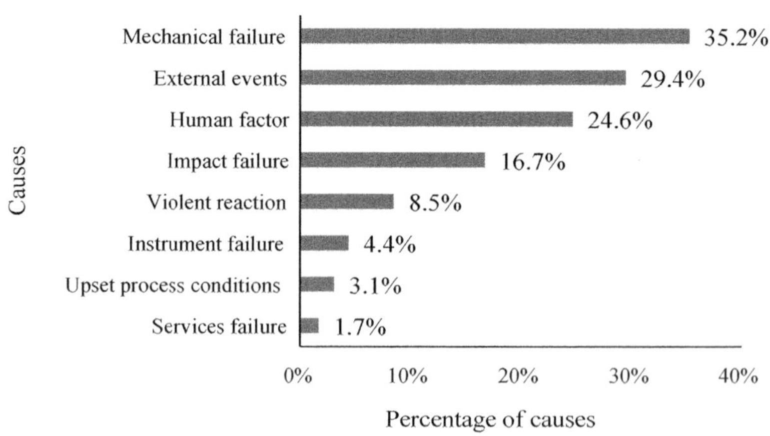

Fig. 1.3 The distribution of domino effect causes

Table 1.4 Comparison between unintentional and intentional domino effects (Chen et al. [1])

Types	Unintentional domino effects	Intentional domino effects
Definition	Domino effects triggered by unintentional events	Domino effects triggered by intentional events
Positions of primary events	Usually occurring at installations	Any positions within chemical plants or outside the area nearby
Sources of hazards	Hazardous materials in chemical installations and hazardous materials form loading and unloading vehicles	Hazardous materials in chemical installations, and external hazardous materials carried by attackers such as explosive devices
Main escalation vectors	Heat radiation, fire impingement, overpressure, and fragments	Heat radiation, fire impingement, overpressure, and fragments
Simultaneous primary scenarios	Usually involving a single installation	Multiple installations can be involved due to multiple target attacks
Protection measures	Safety barriers	Security countermeasures and safety barriers

resulting in unpreventable propagation of major accident scenarios. For instance, water supply infrastructures, power supply systems, communication, and medical facilities may be simultaneously or sequentially attacked by an adversary that has a good knowledge of safety and security in the process industry. Compared with unintentional domino effects, intentional domino effects may be more challenging to prevent since intelligent and strategic adversaries can aim to damage security and security measures. Moreover, adversaries can deliberately induce domino effects by attacking the installations has a high probability of initiating domino effects. Thus both intentional and unintentional domino effects can not be ignored for chemical industrial areas.

According to the failure sequences of installations in domino effects, the initiating events or the primary events are characterized by cardinality 0, whereas cardinality 1 refers to secondary domino events, and cardinality 2 to tertiary domino events, etc. [6]. The propagation from the primary scenario to the secondary scenario may be called first-level propagation, while from a secondary scenario to a tertiary scenario may be called second-level propagation [1]. All the major hazards of fire, explosion, and toxic cloud dispersion can be simultaneously or sequentially present in one disaster due to the propagation of hazardous scenarios. For example, 81% of explosions triggered fires among 330 domino effect accidents [22].

1.4.3 Domino Effect Classifications

In light of these features, domino effects can be classified into several categories according to different criteria. Following the research by Reniers [36], domino effects are divided by eight different criteria, as shown in Table 1.5. These classifications are based on one feature of domino effects. For instance, domino effects can be divided into external domino effects and internal domino effects according to the damaged area of domino effects. External domino effects can only occur in a chemical cluster when multiple chemical plants are involved in a domino effect accident. Besides, A domino effect can belong to different types. If a domino effect accident damages

Table 1.5 Categories of domino events (excluding toxic domino effects) (Chen et al. [1])

Type	categories	Definition
1	Unintentional	Domino effects induced by accidents or natural disasters
	Intentional	Domino effects caused by intentional attacks
2	Fire-induced	The primary event is a fire
	Explosion-induced	The primary event is an explosion
3	Internal	The start and end of the escalation vector characterizing the domino event are situated inside the boundaries of the same chemical plant
	External	The start and end of the escalation vector characterizing the domino event are not situated inside the boundaries of the same chemical plant
4	Direct	Domino events occur as a direct consequence of the previous domino event
	Indirect	Domino events occur as an indirect consequence of a preceding domino event, not being the previous one
5	Temporal	Domino events occur within the same area as the preceding event, but with a delay
	Spatial	Domino events occur outside the area where the preceding event took place
6	Serial	Domino events occur as a consequent link of the only accident chain caused by the preceding event
	Parallel	Domino events occur as one of several simultaneous consequent links of accident chains caused by the preceding event
7	Heat radiation-induced	Domino effects caused by heat radiation
	Overpressure-induced	Domino effect caused by overpressure
	Fragment-induced	Domino effects caused by fragmentation impact
	Coupled	Multiple kinds of escalation vectors are present during the evolution of the domino effect
8	First-level	The highest propagation is first-level order
	Second-level	The highest propagation is second-level order
	Nth-level	The highest propagation is Nth-level order

multiple chemical plants and involves several hazardous scenarios, it can be called an external domino effect or a coupled domino effect.

1.5 Vulnerability Assessment of Installations

Vulnerability assessment can provide reliable models to estimate damage possibility and probability of an asset or a human exposed to an undesired event, which is critical for safety and security risk assessment. The vulnerability assessment for domino effects is to assess the inability of a hazardous installation to withstand the failure induced by escalation vectors. An accident scenario propagates when the generated escalation vectors lead to the failure of hazardous installations. As a result, a vulnerability assessment is essential for assessing the escalation of potential cascading effects. The vulnerability assessment methods in literature can be divided into three categories: deterministic methods, probabilistic methods, and CFD/FEM methods.

1.5.1 Deterministic Methods

Deterministic methods based on propagation thresholds are the earliest vulnerability analysis approaches for assessing the failure of installations exposed to hazardous scenarios. These thresholds can be obtained by experiments and accident analysis [37]. An escalation threshold represents the minimum intensity of physical effects able to cause a propagation, which can be used to develop "rules of thumb" or as a screening tool for the preliminary assessment of possible escalation scenarios [6]. The required minimum distance between two hazardous installations to avoid propagation events is usually called "safety distance" or "effect distance" [4, 13]. Table 1.6 lists escalation thresholds of escalation vectors and safety distances between installations for different equipment.

Escalation thresholds can be transferred to safety distances by consequence analysis [38, 39]. According to escalation thresholds and safety distances, domino effects can be easily assessed while the escalation uncertainties are not considered. The vulnerability of installations depends on the complexity of escalation vectors and the features of target installations. Many uncertainties exist in domino effect escalation since a threshold value derived from a special condition may not suitable for other conditions. As a result, different or contradictory threshold values for domino effect assessment can be observed in the literature and technical specifications [10, 38]. For example, the recommended threshold values and safety distances span over an order of magnitude among different countries in the EU due to the lack of a harmonized approach to assessing major accident hazards in the European countries [24]. Besides, the safety distance obtained by comparing the consequence analysis results of hazardous scenarios with the escalation threshold values may not be conservative.

Table 1.6 A list of escalation thresholds and safety distances (Alileche et al. [24])

Scenario	Escalation vector	Modality	Target category	Escalation threshold	Safety distance
Flash fire	Heat radiation	Fire impingement	Floating roof tanks	Flame envelope	Max. flame distance
–			All other units	Escalation unlikely	–
Fireball	Heat radiation	Flame engulfment	Atmospheric	$H > 100\ \text{kW/m}^2$	Maximum flame distance
			Pressurized	Escalation unlikely	–
		Distant radiation	Atmospheric	$H > 100\ \text{kW/m}^2$	Maximum flame distance
			Pressurized	Escalation unlikely	–
Jet fire	Heat radiation	Fire impingement	All	Flame envelope	–
		Stationary radiation	Atmospheric	$H > 15\ \text{kW/m}^2$	50 m from flame envelope
			Pressurized	$H > 45\ \text{kW/m}^2$	25 m from flame envelope
Pool fire	Heat radiation	Flame engulfment	All	Flame envelope	–
		Stationary radiation	Atmospheric	$H > 15\ \text{kW/m}^2$	50 m from flame envelope
			Pressurized	$H > 45\ \text{kW/m}^2$	25 m from flame envelope
VCE	Overpressure	Blast wave interaction	Atmospheric	$P_S > 22$ kPa	R = 1.75 (ME); 1.50 (BS)
			Pressurized	$P_S > 20$ kPa	R = 2.10 (ME); 1.80 (BS)
			Elongated (toxic)	$P_S > 20$ kPa	R = 2.10 (ME); 1.80 (BS)
			Elongated (flammable)	$P_S > 31$ kPa	R = 1.35 (ME); 0.85 (BS)

(continued)

Table 1.6 (continued)

Scenario	Escalation vector	Modality	Target category	Escalation threshold	Safety distance
	Heat radiation	Fire impingement	Floating roof tanks	Flame envelope	Max. flame distance
			All other units	Escalation unlikely	–
Confined explosion	Overpressure	Blast wave interaction	Atmospheric	$P_S > 22$ kPa	20 m from vent
			Pressurized	$P_S > 20$ kPa	20 m from vent
			Elongated (toxic)	$P_S > 20$ kPa	20 m from vent
			Elongated (flammable)	$P_S > 31$ kPa	20 m from vent
Mechanical explosion	Overpressure	Blast wave interaction	Atmospheric	$P_S > 22$ kPa	R = 1.80
			Pressurized	$P_S > 20$ kPa	R = 2.00
			Elongated (toxic)	$P_S > 20$ kPa	R = 2.00
			Elongated (flammable)	$P_S > 31$ kPa	R = 1.20
	Missile projection		All	Fragment impact	300 m (prob. < 0.05)
BLEVE	Overpressure	Blast wave interaction	Atmospheric	$P_S > 22$ kPa	R = 1.80
			Pressurized	$P_S > 20$ kPa	R = 2.00
			Elongated (toxic)	$P_S > 20$ kPa	R = 2.00
			Elongated (flammable)	$P_S > 31$ kPa	R = 1.20
	Missile projection		All	Fragment impact	300 m ($p <$ 5%)
Point-source explosion		Blast wave interaction	Atmospheric	$P_S > 22$ kPa	–
			Pressurized	$P_S > 20$ kPa	–
			Elongated (toxic)	$P_S > 20$ kPa	–
			Elongated (flammable)	$P_S > 31$ kPa	–

H: heat radiation intensity; P_S: static peak overpressure; R: energy scaled distance; ME: Multi Energy method; BS: Baker-Sthrelow method; p: probability

Although the uncertainty of safety distances derived from thresholds and primary scenarios is inevitable, it is also adopted in inherent safety design [40] and technical specifications [39] due to the simplicity and transparency of the approach.

1.5.2 Probabilistic Methods

To model escalation uncertainties and support quantitative risk assessment (QRA) of domino effects, probabilistic methods are developed to assess the vulnerability of installations. According to probability theory, the vulnerability of an installation exposed to an escalation vector can be modeled by a distribution function such as a probit function and normal distribution. Probit analysis is a well-known method to evaluate the dose–effect relationship for human responses to toxic substances, thermal radiation, and overpressure [41], as shown in Eq. (1.1):

$$Y = a + b\ln(D) \tag{1.1}$$

D denotes the "dose" and Y is the probit value representing the "response". a and b are constant parameters, depending on the type of dose and response. Y characterizes the vulnerability of humans exposed to toxic substances. Then the death or injury probability can be obtained using a cumulative standard normal distribution, as shown in Eq. (1.2) [11].

$$P_r = \Phi(Y - 5) \tag{1.2}$$

Φ is the function of a cumulative standard normal distribution, P_r is the death or injury probability. The probit model (Eqs. (1.1) and (1.2)) can also be extended to assess the vulnerability of installations exposed to the most common escalation vectors in domino effects: overpressure, heat radiation, and fragments.

(1) **Overpressure**

Overpressure and fragments are the two main escalation vectors induced by explosions. Bagster and Pitblado [10] proposed a probability approach defining a damage probability function based on the distance from the center of the primary scenario and the safety distance. Khan and Abbasi [34] adopted a probit function approach to model the damage probability caused by overpressure, considering peak overpressure (static pressure) and dynamic pressure. The probit function for overpressure is obtained by using overpressure as a substitute for the dose D in Eq. (1.1). However, using a general probit model for assessing the vulnerability of all vessels exposed to overpressure may lead to large deviations. Cozzani and Salzano [42] thus developed a probit model for each type of vessel to improve the performance of the probit model. The vessels are divided into four categories: atmospheric, pressurized, elongated, and small. The equipment-specific models can significantly reduce the errors caused by the general probit model, presenting an important difference between the damage probabilities and the damage threshold of different categories of vessels.

To reduce the regression errors of probit models, the equipment-specific models were improved by distinguishing the extent of damage and assigning a linear relationship between the probit value and the observed thresholds for each type of damage [43]. In that case, a series of more accurate models were obtained, decreasing the errors between observed data and predicted data. Besides, Mukhim et al. [3] further improved the work of Zhang and Jiang [43] by classifying the equipment into 11 categories and developing a probit model for each category of equipment. These vulnerability models for overpressure can also be applied to vulnerability assessment for installations subject to explosives [44]. Furthermore, the probit models were coupled to simplified calculation models for peak overpressure to develop a straightforward approach for estimating safety distances caused by blast waves and damage probability as a function of the scaled distance. Besides, Salzano and Cozzani [45] studied the intensity of the loss of containment following overpressure wave interaction with process equipment using fuzzy set analysis, which may be used to assess second-level escalation. Although these probit models can address the uncertainties in the escalation caused by overpressure, some important parameters that have essential impacts on vulnerability are neglected, such as the thickness of the vessel wall and vessel materials.

(2) **Fragments**

Fragments and missile hazards may be induced by several accident scenarios such as BLEVE, physical explosions, confined explosions and runaway reactions, etc. Among these accident scenarios, the BLEVE is responsible for most industrial accidents involving fragment projection and usually leads to severe consequences [46]. Gubinelli et al. [47] proposed a probabilistic model according to the event sequence to assess the damage probability induced by fragments. In the model, the damage process caused by fragments are divided into four independent events, and the damage probability can be obtained by multiplying the probability of each event:

$$f_{d,F} = f_p \times P_{gen,F} \times P_{imp,F} \times P_{dam,F} \quad (1.3)$$

$f_{d,F}$ denotes the damage probability induced by a fragment F; f_p represents the probability of primary event; $P_{gen,F}$ denotes the probability of the fragment to be generated in the primary event; $P_{imp,F}$ represents the probability of impact between the fragment and a target installation; $P_{dam,F}$ represents the probability of target damage given the impact with the fragment. As a result, the total damage probability can be represented as the sum of the probability caused by each fragment [47]. This model characterizes the impact of the likelihood of sequential events (where regarded as independent events) on the damage of an installation caused by fragments, while simultaneous damages induced by multiple fragments on one installation are ignored. Besides, the dependencies among the sequential events are not considered. Projected fragments can generate secondary accidents at relevant distances from the primary scenario due to possible large projection distances. The hazards associated with projected fragments are related to the number of fragments, fragment mass, and velocity. To assess the vulnerability of installations subject to fragments, the following steps are usually adopted: (i) calculation of explosion energy, (ii) estimation of the number of fragments, (iii) calculation of initial velocity, (iv) calculation of the angle of departure, and (v) calculation of trajectory [48]. Because it is very difficult to accurately obtain the initial velocity and the departure angle, thus thresholds for fragments are rare in the literature. To reduce the complexity of probability calculation and consider more uncertainties, Monte Carlo simulation and probability density functions are always used to simulate the damage process of fragment projection [46, 48]. Zhang and Chen [49] derived a formula for the initial projection velocity of fragments by taking the explosion moment as a polytropic process and solving the energy transformation equation.

(3) **Heat radiation**

Different from the damage caused by an explosion, the damage mechanism of fire is a dynamic process, i.e., installations exposed to heat radiation do not fail immediately. As time goes by, the vulnerability of exposed vessels increases due to the build-up of temperature/pressure inside the vessel. The vessel fails when a loss of containment emerges. The delay time between the start of the fire and the failure of the target equipment is named "time to failure" (*ttf*) [37]. The concept of *ttf* is derived for assessing first-level domino effects, neglecting the time-lapse in higher-level escalations. To apply it in higher-level propagations, the concept of "residual time to failure" (RTF) is proposed by Chen et al. [15]. The vulnerability of a vessel can be dynamically assessed by updating RTF when the received heat radiation changes in higher-level propagations. In terms of a special industrial area, the two assumptions may be adjusted to obtain more accurate results. Therefore, emergency response has a huge impact on the vulnerability of installations besides heat radiation intensity (threshold). In this context, Landucci et al. [50] developed a probit model for estimating the damage probability of storage tanks exposed to fire based on the *ttf* obtained via empirical formulas. The probit model is based on three assumptions: (i) 10% probability of failure for $ttf = 5$ min, which is equal to the minimum time required to start on-site emergency response operations; (ii) 90% probability of failure for $ttf = 20$ min, which is equal to the minimum time required to start the mitigation actions. Chen et al. [9] extended this work to overcome the limitation of the "probit model" approach in higher-level propagations, modeling the uncertainty of emergency response using a normal distribution.

1.5.3 CFD/FEM Methods

In recent years, more attention has been paid to advanced numerical methods such as Computational Fluid Dynamics (CFD) and the Finite Element Method (FEM) due to their strengths in physical effect simulation. These advanced methods are considered to be a promising tool to support the assessment of complex accidental scenarios, such as three-dimensional pool fire and vapor cloud dispersion [51]. Landucci et al. [50] modeled the failure of storage tanks exposed to fire using a commercial FEM code. The FEM model can simulate the thermal and mechanical parameters of vessel shells under heat radiation, such as heat radiation, wall temperature, and stress. A storage vessel is assumed to fail when the equivalent intensity of combined stress becomes greater than the maximum allowable stress. This work was extended to model the performance of different materials proposed for the passive fire protection of tanks [52]. Jujuly et al. [53] developed a three-dimensional computational fluid dynamics (CFD) simulation of a liquefied natural gas (LNG) pool fire using ANSYS CFX. In this study, shell temperature and heat radiation thresholds were used to determine the failure of storage vessels. The results show that wind speed has a significant contribution to the behavior of a pool fire and its possible accompanying domino effects. Besides, ANSYS FLUENT was also used to model the heat transfer and pressure build-up in LPG vessels exposed to fires [51]. FLACS software developed by Gexcon AS was also be used to model flammable cloud dispersion and VCE explosion, supporting domino effect assessment [54]. These studies indicate that using advanced simulation tools can obtain a more precise assessment of failure conditions of vessels engulfed in fires, thus supporting the development of vulnerability models for process equipment exposed to fire. More recently, the Fire Dynamics Simulator (FDS) is adopted to simulate tank and dike pool fires in a tank farm [55]. CFD simulation may obtain more accurate results of physical effects and thus facilitate vulnerability assessment of installations exposed to escalation vectors, but it is very complex, time-consuming, and costly [56]. With the rapid development of computer science, applying CFD methods in vulnerability assessment may become more accessible and acceptable for researchers and engineers in the future.

1.6 Risk Assessment of Domino Effects

Based on the vulnerability assessment methods reviewed in Sect. 1.5, a domino effect assessment can be conducted to obtain the likelihoods and consequences of domino effect scenarios. This section reviews past research on the risk assessment of domino effects. The risk assessment methods used for domino effects are divided into three categories: analytical methods, graphical methods, and simulation methods.

1.6.1 Analytical Methods

Analytic methods refer to the methods using exact theorems to present formulas that can be used to obtain domino effect risks without using graphic and numerical methods. Bagster and Pitblado [10] developed a program based on a squared decay function, considering higher-level escalations and multi-vector. In this study, the differences of physical mechanisms among different escalation scenarios were not fully addressed since safety distance was used for all escalation scenarios.

Different from the safety distance-based approach, probabilistic methods can be more suitable to model uncertainties. Khan and Abbasi [57] established a probabilistic method to assess domino effects and developed a software package called MAXCRED based on the assessment method. In this study, a maximum credible accident analysis is conducted to assess the potential consequences of a chemical plant, and Probit models were used to assess the vulnerability of installations exposed to different escalation scenarios. The results show that a confined vapor cloud explosion followed by a pool fire would be the worst scenario with the maximum likelihood of triggering domino effects. Khan and Abbasi [30] updated the work and developed a new user-friendly software (DOMIFFECT). This software adopted an analytical approach to model different escalation scenarios (i.e., fire, explosion, and toxic release) and the interactive impacts between fire and explosion. The updated software was used to estimate possible hazards from loss of containment to explosions, deal with interactions among different escalation scenarios, assess the probability of domino effect scenarios, and estimate the potential consequences. Furtherly, Khan et al. [58] the DOMIFFECT software was regarded as a consequence analysis module of a risk analysis methodology called Optimal Risk Analysis (ORA).

Cozzani and Salzano [42] established a quantitative assessment methodology for overpressure-induced domino effects using equipment-specific probit models. This study demonstrated that individual risk increases up to an order of magnitude by considering domino effects. Following this work, Cozzani et al. [11] developed a systematic quantitative risk assessment method for domino effects. In this QRA method, the vulnerability of installations exposed to escalation vectors of heat radiation and overpressure were modeled by probit models, while that of fragment projection is characterized by a probabilistic model. According to the methodology, the individual risk, societal risk can be obtained considering all the credible combinations of secondary events. Besides, the QRA methodology was integrated with GIS software called Aripar-GIS [59]. By applying this software in chemical plants, possible escalation targets can be automatically identified, determining individual and societal risk induced by possible domino effects. Combining the vulnerability assessment methods for installations subject to natural disasters, the QRA framework was extended to assess domino effects caused by natural disasters [17].

Besides, TNO (Netherlands Organization for applied scientific research) developed a QRA tool for industrial plants with respect to dust explosion hazards, considering the first-level escalation based on safety distances [60]. To quantify domino effect risk, Zhang and Chen [61] proposed a QRA method based on failure mechanism analysis using a visualized risk cloud figure. Kadri et al. [62] established a QRA method for domino effect induced by fire and explosion in hazardous material storage areas by defining a concept of “domino system”. Zhou and Reniers [63] introduced a matrix-based approach for quantitative risk assessment of fire-induced domino effects, considering possible synergistic effects. Besides, commercial QRA or consequence software may also facilitate domino effect assessment, such as FLACS developed by Gexcon AS, Shepherd and FRED developed by Shell as well as EFFECTS and RISKCURVES developed by TNO [1].

With the construction of parallel pipelines, increasing attention in the scientific literature has been paid to domino effects induced by parallel pipelines [64]. Compared with domino effects in industrial plants, domino effect management with respect to parallel pipelines may be more difficult since they are over long distances [1]. Ramirez-Camacho et al. [64] established an analytical model to assess the likelihood of domino effects in parallel pipelines. The likelihood was a function of the location of the hole, the jet direction and solid angle, the diameters of pipelines, and the distance between pipelines. In terms of underground parallel pipelines, Silva et al. [65] developed an analytical model based on historical accident data and pipeline crater models, indicating that a separation distance of 10 m would be sufficient to prevent domino effects caused by parallel pipelines.

1.6.2 Graphical Methods

Chemical industrial areas consist of various hazardous installations with different domino effect potentials. Some installations exhibit a high probability of initiating domino accidents, while other installations are more likely to propagate domino events. These installations can be regarded as nodes. The quantitative possibility of accident propagation may be represented by the weight of the links between nodes in a network or graph [9, 13, 66]. Compared with analytical methods, graphical models may provide a framework for the evolution of domino effects, tackling complex domino scenarios and higher-order propagations.

(1) **Graph/network models**

Reniers and Dullaert [13] proposed a domino effect assessment method in chemical industrial plants based on graph theory. In a graph, nodes represent chemical installations, and arcs between each pair of nodes denote escalation vectors. The weight of each arc denotes the escalation likelihood from a tail installation to a head installation. In this study, a distance-based matrix called Domino Danger Unites Matrix (DDU) was defined as the weight of arcs, characterizing possible accident scenarios from one installation to another [13]. By the graphic approach, hazardous installations in an industrial area can be modeled as a whole in terms of the danger they pose to each other. As a result, the developed algorithm and software critical installations with a high probability of initiating or propagating domino effects can be identified, supporting escalation prevention decision-making [13].

(2) **Graph metrics**

Khakzad and Reniers [66] analyzed the vulnerability of process plants in the context of domino effects using graph metrics such as betweenness, out-closeness, and in-closeness of directed graphs, and closeness of undirected graphs. The betweenness of a node represents the fraction of geodesic distances or the weights of edges between all pairs of nodes that traverse the node of interest. As a result, the betweenness can represent both the escalation likelihood and the damage likelihood of installations. The out-closeness of a node represents the number of steps needed to reach every other node of the graph, indicating the escalation potential of the node (installation). Finally, the in-closeness of a node denotes the number of steps needed to reach the node from every other node of the graph, reflecting the likelihood of nodes (installations) being damaged by domino effects. Applying these metrics, critical installations or the most vulnerable installations in an industrial area can be quickly identified [66].

(3) **Bayesian network**

In risk assessment and artificial intelligence, Bayesian network (BN) is a widely used probabilistic graphical tool to model uncertain knowledge and dependency in probabilistic systems [67, 68]. Khakzad et al. [69] firstly introduced a BN to model the domino effect and estimate the domino effect probability at different escalation levels. By the structure of the network, possible synergistic effects can be considered, considering the complex interaction and conditional dependencies among the units involved in the domino effects. In that case, limited assumptions such as independent events or random or binomial selection of target units can be relaxed [69]. A Bayesian network represents one evolution path of domino effects, while the evolution path is uncertain at the beginning of domino effects. Similar to the approach based on graph metrics, the Bayesian network approach may be used to model the conditional dependencies of a critical evolution path. If the uncertainties of evolution paths need to be considered, multiple Bayesian networks need to be developed [70]. Figure 1.4 is a case of a dynamic network for domino effect assessment. In this network, the modeled evolution path is determined (T1 $\rightarrow$ T3 $\rightarrow$ T2 and T4 $\rightarrow$ T5).

Based on the ordinary Bayesian network approach [69], Khakzad [71] developed a dynamic Bayesian network (DBN) model considering both the spatial and temporal escalation of domino effects. The DBN explicitly models time dependencies and identifies the most probable sequence of accidents. Compared to the ordinary BN, this method can reflect the characteristics of a domino effect much better than the

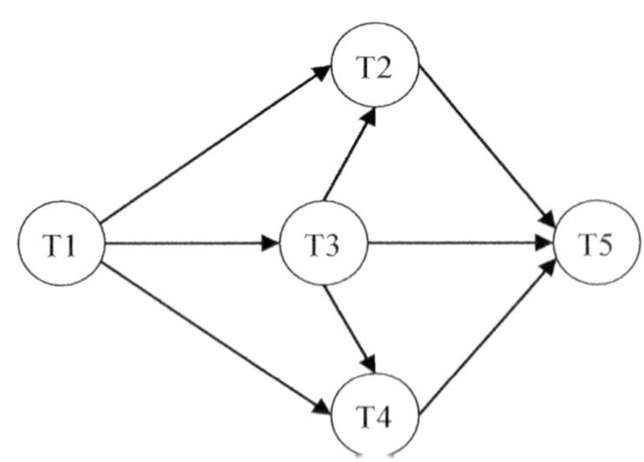

Fig. 1.4 A possible propagation pattern of the domino effect represented by BN (Chen et al. [1])

most probable combination of accidents. Similar to the ordinary BN approach, only the most probable sequence of accidents is considered to determine the structure of BN. Besides, DBN was employed to analyze domino effect escalation in wildland-industrial interfaces [72] and the escalation effects caused by intentional attacks [16]. Graph metrics can rapidly identify the critical installations and support the establishment of a critical evolution path, while BN can be used to calculate domino effect probabilities based on the critical evolution path.

(4) **Dynamic graphs**

Graph metrics is a threshold-based approach that neglects the uncertainties and time-dependencies in escalation assessment. To overcome these limitations and quantify domino effect risks, Chen et al. [9] developed an approach based on dynamic graphs to model the spatial–temporal evolution of domino accidents. This approach models the uncertainties and time-dependencies in escalation assessment and overcomes the limitation of the "probit model" w.r.t higher-level escalation. Complex spatial evolutions such as synergistic effects and parallel effects and the superimposed effects in spatial evolution are considered in this approach. Different from static graph models, a dynamic graph consists of sequential static graphs, and each static graph represents an escalation status at a specific time. Therefore, a dynamic graph is updated with the evolution of domino effects. Compared to the static graph, which provides merely a snapshot of the whole process at once, dynamic graphs seem to be able to model the dynamic evolution of domino effects (escalation sequence) [15]. Figure 1.5 shows a comparison between a static graph and a dynamic graph. The static graph (a) only provides one snapshot of the entire evolution process, while the dynamic graph (b) models the dynamic evolution process. By the application of the dynamic

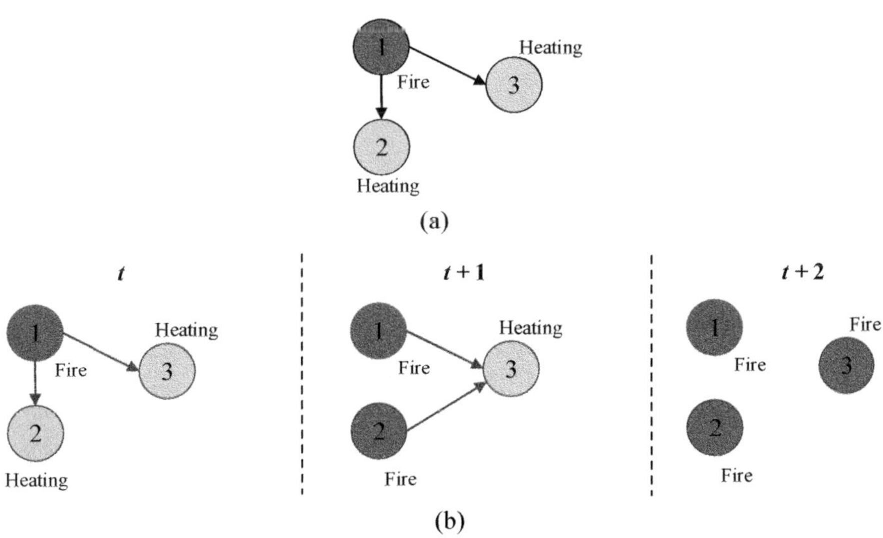

Fig. 1.5 The structure of (**a**) a static graph model and (**b**) a dynamic graph model of domino effects

graph method, we can rapidly obtain the evolution time, evolution paths, the status of installations, the failure probabilities of installations. Besides, possible multiple failures of installations within a primary scenario caused by intentional attacks or natural disasters can also be addressed using the dynamic graph approach. Furthermore, combining this approach with Monte Carlo simulation, the uncertainties in scenario evolution and escalation assessment can also be addressed [73].

(5) **Petri-net models**

Petri-nets are graphic tools that are widely used to analyze and simulate concurrent systems [74]. A Petri-met consists of two sets of nodes: the set of places representing system objects, and the set of events or transitions denotes the dynamics of the system. Zhou and Reniers [75] proposed a Petri-net model to analyze domino effect evolution induced by overpressure. The probabilistic dependencies in different events were modeled by the token of a place with a probability value. Similar to the BN method, the structure of the Petri-net is developed based on threshold values. Then the probability calculation is performed according to the developed network [75]. Besides, Petri-nets were applied to model the complex interaction and time dependencies among installations during the evolution of fire-induced domino effects [76].

(6) **Event tree method**

Event tree analysis is a forward, top-down, logical modeling technique using dichotomous conditions to model the propagation of events by different branches of the tree [77]. It is widely used to analyze the sequential events following an initial event. Alileche et al. [78] developed a methodology based on event tree analysis to model the propagation of domino effects. By applying the event tree analysis, different accident scenarios caused by a loss of containment can be identified to support propagation analysis.

1.6.3 Simulation Methods

Simulation methods are used to generate possible scenarios and then obtain the probabilities by the statistics of each scenario. It can simplify the complexity of analytical techniques and thus improve the efficiency of tracking complex accident evolutions. Abdolhamidzadeh et al. [23] proposed a Monte Carlo method for assessing the probability of domino effects and the failure frequency of installations. Then, this simulation method was extended to model multi-scenario and higher-level escalations [79]. The Monte Carlo-based simulation approach can successfully model the spatial evolution of domino accidents while it may be time-consuming. Similar to the BN approaches, it is a purely probabilistic tool based on randomly generated numbers, lacking actual accident propagation mechanisms. Besides, Zhang et al. [80] proposed an agent-based simulation tool considering installations' states that was developed to analyze the second or higher-level propagations and temporal dependencies. An algorithm based on Monte Carlo simulation is used to solve the model. As a result, it takes enormous computation time for realistic chemical clusters with many installations.

1.7 Safety and Security Management of Domino Effects

Safety and security management in the chemical and process industry reduces the hazard of a process, the likelihood of accidents, and the consequences. A wide variety of strategies, techniques, procedures, policies and systems are used to this end [81]. Since domino effects can be triggered by both intentional and unintentional events, Reniers et al. [14] deemed that both safety and security are important for domino effect management. Therefore, a wide variety of safety and security measures are needed to prevent or mitigate possible intentional domino effects and unintentional domino effects. According to Bollinger and Crowl [81], risk reduction strategies can be divided into four categories: inherent, passive, active, and procedural. In this review, strategies for managing domino effects are divided into four categories: inherent safety, safety barrier management, cooperative prevention strategies, and security strategies for intentional domino effects.

1.7.1 Inherent Safety

Inherent safety aims to remove the hazard at the source rather than to accept the hazard by attempting to mitigate the effects. Inherent safety is a practical and straightforward approach that has received the most attention in the prior development of assessment tools [82]. Kletz [83] proposed five inherent safety principles: intensification or minimization, substitution, moderation by attenuation, simplification, and moderation by limitation of effects, as shown in Table 1.7.

Some of the principles may be difficult to implement in practice, and not all these principles can be used in domino effect management. These principles can always be used to identify inherent safety measures for escalation prevention if they are considered in an assessment. Inherent safety strategies can prevent the initiation of the accident, decrease the potential of domino effects or terminate the accident sequence. Past domino effect management work related to inherent safety, includes developing inherent safety indexes, optimizing layout, and optimizing inventories.

(1) **Inherent safety indexes**

Cozzani et al. [84] analyzed possible escalation scenarios to identify inherent safety measures related to the prevention and mitigation of domino effects. This study shows that the principle of "limitation of effects" is more effective, and integrating inherent safety criteria with passive or active protections may be a promising route for domino

Table 1.7 Inherently safer principles (Chen et al. [1])

Inherently safer principles	Description
Intensification or minimization	Using so little of the hazardous material that there is no significant risk if it all leaks out
Substitution	using a less hazardous material or a process that is less likely to develop into an accident scenario (e.g., a runaway reaction)
Moderation by attenuation	Using the hazardous material in the least hazardous form (under the least hazardous condition)
Simplification	Using simpler plants that provide fewer opportunities for error and less equipment that can fail
Moderation by limitation of effects	Duplicate or separate processes, critical activities, installations, to limit the possibility of severe effects

effect management in chemical and process plants. According to these inherent safety principles, Cozzani et al. [40] defined a set of inherent safety indexes to identify process and layout hazards related to escalation events. The hazard indexes can be used to identify inherent measures for domino effect management, such as measures in layout design. Tugnoli et al. [85] examined the five inherent safety principles and demonstrated that the principles of "attenuation", "simplification", and "limitation of effects" are practical for layout design. According to these principles, they established indexes for plant layout design, considering possible escalation scenarios [86]. To compare different technologies for hydrogen storage, Landucci et al. [87] developed a set of inherent safety key performance indicators according to consequence assessment and credit factors of possible LOC events. These studies evidence that the application of inherent safety principles in domino effect management can effectively prevent and mitigate domino effects.

(2) **Layout optimization**

A widely used inherent safety practice in domino effect management is layout optimization for chemical plant layout design. Lee et al. [88] proposed a nonlinear program to optimize the distribution of explosive facilities, considering possible escalation effects caused by fire, overpressure, and fragments. The optimization can minimize the total escalation probabilities caused by different escalation scenarios, ignoring the difference of likelihoods of different primary scenarios. The developed computer-aided module (MiniFFECT) sequentially allocates hazardous installations in a limited land [89]. This optimization approach was further improved to obtain the optimal plant layout minimizing the total weighted consequences of different escalation scenarios [90]. Besides, the research was also extended to minimize the total construction costs, including pipeline connection cost, protection cost, and land-use cost [91].

Jung et al. [92] developed an optimization approach for facility siting and layout decision-making using a mixed-integer nonlinear program (MINLP). MINLP is a widely used algorithm used for optimization problems with continuous and discrete variables and nonlinear functions in the objective function and/or the constraints. This study aimed at minimizing the overall cost, including land costs, interconnection costs between facilities, and risk costs derived from possible structural damage caused by overpressures). The optimization results can be used to safely and economically determine the location of new facilities. Based on probit models and consequence indexes, the MINLP approach was also used to optimize chemical plant layout and thus to reduce domino effects [93]. Besides, Latifi et al. [94] proposed an MINLP formulation to optimize process plant layout, considering major accidents such as toxic release, fire, explosion, and possible domino effects induced by them.

Bernechea and Arnaldos [95] proposed a multi-objective optimization method to optimize the design of storage facilities based on inherent safety design and quantitative risk assessment. Multi-objective optimization is a decision-making tool using multiple criteria in which two or more objectives subjected to different restrictions are simultaneously optimized. Consequently, it can be used to balance the conflict between domino effect management and the reduction of investment costs [95]. Nomen et al. [96] proposed a plant layout design method based on QRA results. In this approach, a simple criterion based on the surface enclosed in isorisk curves is used for comparing different QRA results. Khakzad and Reniers [97] developed a multi-criteria decision analysis (MCDA) tool using an analytic hierarchical process (AHP) to support chemical plant design. In this study, BN is combined with MCDA techniques for layout optimization, considering a variety of hazardous installations and multiple accident scenarios.

(3) **Inventory optimization**

Besides plant layout, domino effects also depend on the distribution of hazardous substances. Thus optimization of the distribution of chemical inventories is a practical option when reducing the total mass of hazardous substances is impossible. Khakzad et al. [98] proposed an approach based on DBN to optimize the chemical plant inventory, minimizing domino effect risk. This study indicated that optimization of the distribution of hazardous inventories is essential for process plants with fixed safety distances.

1.7.2 Management of Safety Barriers

Passive barriers in process safety management refer to any measures that can reduce either the frequency or consequence of the hazard without the active functioning of any device, such as dikes, firewalls, and fireproofing coatings. Active barriers are any measures that can detect and respond to process deviation from normal operation using controls, alarms, safety instrumented systems or functions, and mitigation systems, such as water deluge systems (WDS), emergency shutdown systems (ESD), and emergency depressurization systems (EDP) [1, 15]. Previous work on safety barriers related to domino effects is divided into three categories: barrier performance assessment, barrier optimization, and emergency response.

(1) Safety barrier performance

The performance of a safety barrier for preventing escalation prevention depends on the "availability" and "effectiveness" of the barrier. The availability represents the complement of the probability of failure on demand (PFD) of the safety barrier, and the effectiveness refers to the conditional probability of the escalation being prevented given the barrier is activated. Landucci et al. [99] proposed a quantitative method for assessing the performance of safety barriers. A fault tree was used to quantitatively assess the PFD of safety barriers, and an event tree is adopted to analyze domino effects, considering the function of relevant safety barriers. The performance of different barriers can be obtained by combining the quantitative method with key performance indicator analysis [100]. Table 1.8 lists the PFD values of the common-used safety barriers in the chemical and process industry.

The performance of safety barriers strongly depends on external factors such as external temperature, wind and wave height, etc. [101]. Consequently, harsh environments possibly reduce the performance of safety barriers and the availability of emergency resources [101]. Harsh environments refer to those climatic conditions that may be difficult for people to work and for equipment to be normal operation [102]. For instance, the reliability of emergency response procedures may be decreased due to harsh environments, resulting in the delay of escalation control and the non-achievement of the optimal rescue time. To address the uncertainty

Table 1.8 PFD values of common-used safety barriers (Chen et al. [1])

Safety barrier	Actuation type	Proportioning method	PFD
Foam-water sprinkler system	Pneumatic	In-line educator[a]	5.43×10^{-3}
		Metering proportioning[b]	5.01×10^{-3}
		Bladder tank[c]	3.76×10^{-3}
	Electric	In-line educator	5.39×10^{-3}
		Metering proportioning	4.96×10^{-3}
		Bladder tank	3.72×10^{-3}
WDS for LPG vessels protection	Pneumatic	–	1.89×10^{-2}
	Electric	–	4.33×10^{-2}
WDS for horizontal separator protection	Pneumatic	–	2.24×10^{-2}
	Electric	–	2.24×10^{-2}
ESD system	–	–	3.72×10^{-4}
Pressure Safety Valve (PSV)	–	–	1.00×10^{-2}
Fireproofing coating	–	–	1.00×10^{-3}
Emergency intervention	–	–	1.00×10^{-1}

[a] In line educator has an inlet pressure which can be brought a significant distance from the engine
[b] Metering proportioning is located in line to the front brakes to allow pressure to cause the valve to open allowing pressurized brake fluid to flow to the caliper
[c] Bladder tank includes a pressure-rated tank with an internal elastomeric bladder for foam concentrate storage

and complexity induced by harsh environments, Landucci et al. [101] developed an event tree approach to address the influence of harsh environmental conditions on the performance of hardware safety barriers. Besides, Khakzad et al. [103] established a dynamic Bayesian network to address the degradation of safety barriers during domino events. In this study, an exponential probability distribution was used to model the availability of safety barriers and time-dependent fragility models were adopted to assess the failure probability of installations exposed to fire.

(2) Optimization of safety barriers

Performance assessment can identify effective safety barriers for preventing and mitigating escalation effects such as fireproof coatings and WDS. However, these safety barriers may not be implemented for all equipment in a chemical industrial area due to the safety budget. Therefore, a risk-based methodology is developed to identify fireproofing zones in the initial phases of layout definition, considering domino effects caused by both pool fire and jet fire [104]. This study used a risk matrix to rank the severity of different LOC scenarios and thus identify the reference scenarios. Then, plotted envelopes corresponding to the reference LOCs were used to identify the fireproof zones [104]. Besides, Ghasemi and Nourai [105] developed an approach to determine the water application rate for protecting storage tanks from an external non-contacting fire. The optimization can lead to at least 25% saving in a tank farm area by calculating the water application rate to reduce the separation distance between adjacent tanks.

Janssens et al. [106] developed an optimization method to allocate safety barriers to mitigate the consequences of possible fire-induced domino effects. The optimization aimed to maximize the total failure time associated with a domino effect given a limited budget. This study considered the decision-making on the allocation of safety barriers a knapsack problem and proposed a metaheuristic algorithm to obtain the optimal allocation strategy. This approach can support the allocation of safety barriers under a limited budget. Khakzad et al. [107] applied a cost-effective analysis for decision-making on the allocation of safety barriers. Cost-effectiveness analysis does not strictly require the monetization of protection benefits while always needs to compute cost-effectiveness ratios (CERs). CERs are used to select the most effective protection strategies between two options [5]. The cost-effective analysis based on the limited memory influence diagram can identify the most cost-effective strategy for the allocation of safety barriers.

(3) Emergency response

In the process and chemical industry, emergency response actions are essential for protecting installations, the public, workers, and the environment [1]. Besides, emergency response actions also impact the evolution of accidents and thus play an important role in domino effect management. However, the performance of emergency response is difficult to assess due to the uncertainties related to emergency response procedures [15]. For instance, the performance of firefighting depends on the skills and preparedness of emergency responders, the number of firefighting trucks, the distance between the water resources and the chemical plant, and the time required to start the emergency operations [108]. In terms of intentional attacks on chemical facilities, more than one fire may co-occur, and these fires may lead to domino effects at different locations in a chemical industrial area. In that case, emergency management may be challenging since it is challenging to allocate limited emergency resources to multiple locations. In light of these challenges, much work on emergency management has been done in recent years.

Zhou and Reniers [109] studied the emergency response against simultaneous large-scale fires using a Petri-net simulation method. The simulation analyzed the executing actions and the system status during an emergency response process to obtain the optimal strategy for the allocation of firefighters. The study shows that the allocation of firefighters should be based on fire severity rather than average distribution; the effects of backups on firefighters depend on particular fire scenarios; thus it is not always necessary to improve the backups [109]. A further analysis based on timed colored hybrid Petri-net (TCHPN) indicates that cooling adjacent tanks is more important for preventing fire-induced domino effects [110]. This work was improved to deduce the consequence-antecedent relationship between an accident and the emergency response actions using a fuzzy Petri-net [111]. Besides, an approach combining an event sequence diagram and Monte Carlo simulation was developed to assess emergency response actions considering sequence, duration, correctness, and mutual interaction [112].

Khakzad [113] developed a risk-informed approach based on DBN for emergency response analysis in oil terminals. This study can identify optimal firefighting strategies, especially when the number of fire trucks is insufficient to handle all the vessels exposed to fires. The study shows that cooling an exposed vessel can immediately reduce the likelihood of fire escalation, while suppressing a burning vessel can not quickly reduce the emitting heat radiation. The results from a graph-based approach indicate that suppression and cooling of tanks with the highest out-closeness index is an optimum firefighting strategy [114]. Besides the firefighting optimization based on risk reduction, Cincotta et al. [108] proposed a new optimization based on the resilience concept. This study is used to optimize firefighting strategies, maximizing the resilience of process plants. The failure probability of installations was considered in the developed resilience metrics, whereas the recovery ability was ignored. Hosseinnia et al. [115] developed an emergency response decision matrix to tackle domino effects in chemical clusters. A decision tree of emergency levels and an alert notification system based on a decision matrix are used in the methodology. Compared with unintentional domino effects, the emergency management for intentional domino effects needs to consider the security forces. In target identification, the attractiveness of installations to adversaries should be addressed. Also, the consequence analysis should consider the vulnerability of installations exposed to intentional attacks and the vulnerability of installations subject to escalation effects [115].

1.7.3 Cooperative Prevention

A chemical industrial park or so-called chemical cluster always consists of multiple chemical plants belonging to different companies. The implementation of safety and security resources in one chemical plant may benefit nearby plants due to the prevention and mitigation of possible external domino effects. From a safety viewpoint, domino effect scenarios in the cluster are unknown to other plants if there is no collaboration. Moreover, security measures may increase the (security) risk of nearby plants since the attractiveness of the chemical plants for possible common adversaries may decrease with the implementation of protection measures. As a result, cooperative prevention is thus needed to prevent and mitigate domino effects in chemical clusters.

(1) **Cooperative prevention of external domino effects**

In a chemical cluster, a domino effect can be an internal domino effect or an external domino effect. Internal domino effects refer to the escalation effects occurring within the boundaries of the plant, while external domino effects are the escalation effects that propagate outside the boundaries of the plant [12]. Due to possible external domino effects, a terrorist attack on chemical plant A may have an impact on company B, resulting in a major catastrophic disaster in the cluster. These scenarios may be more likely to unfold if the terrorist has access to sufficient and accurate information about domino effects. These catastrophes may damage multiple plants in a chemical cluster, and it is therefore important to involve all plants in the chemical cluster to collaborate in cross-plant safety [116]. Even if a company invested in preventing domino effects while its neighbors have not, the company might suffer from major accidents triggered by an initial event in an adjacent chemical enterprise in the chemical cluster. Besides, the neighboring company may become more attractive for terrorists since its attractiveness increases if no security investments are implemented [116].

To manage possible external domino effects, Reniers et al. [31] examined risk analysis tools used by 24 chemical plants in Belgium. This work identified the at the time of the study used practices in the chemical industry subject to European Seveso legislation and examined how the risk analysis approaches may be integrated to improve the safety policy. The survey shows that the exchange of expertise and cooperation can lead to a safer working environment. Three risk analysis tools (HAZOP, what-if analysis, and the risk matrix method) are identified as promising tools for stimulating inter-company cooperation. Based on the investigation, an external domino effect management framework (Hazwim) based on identified risk analysis methods was established. The Hazwim framework consists of a process scheme and an organized schedule, which is very useful for decision-making to prevent external domino effects [12]. Besides, Reniers et al. [117] applied a cooperative game to model the decision making of different plants within a chemical cluster on prevention investments.

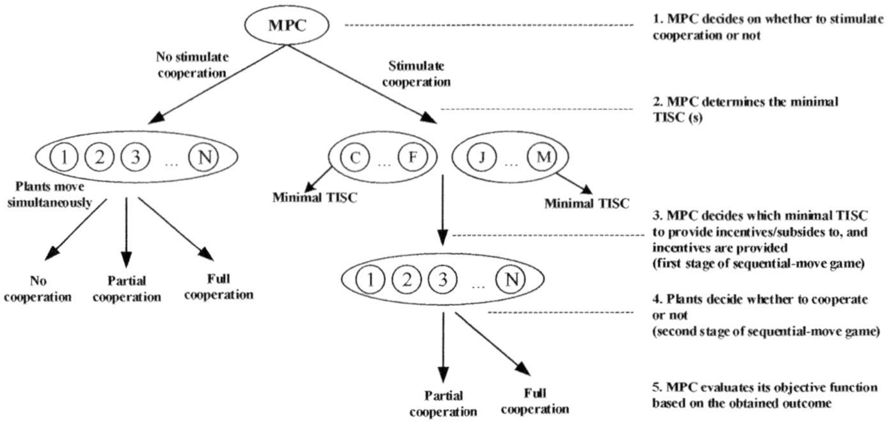

Fig. 1.6 The decision procedures of a two-stage sequential move game (Pavlova and Reniers [118])

According to the approach, a win–win strategy or so-called Nash equilibrium where all the players (companies) can obtain benefits from the investment in prevention measures.

(2) **Enhancing safety and security cooperation**

Due to the extremely low probabilities of external domino effects, trust and confidentiality concerns, it may be difficult to obtain an agreement among companies in the investment in cross-plant preventive measures [36]. Pavlova and Reniers [118] thus developed a sequential-move game to enhance safety and security cooperation within chemical clusters dealing with domino effects, as shown in Fig. 1.6.

As shown in Fig. 1.6, an institution called Multi-Plant Council (MPC) could be established to stimulate the prevention cooperation in a chemical cluster [118]. The MPC would be responsible for a continuous follow-up of safety and security improvements at different companies in the cluster. In the two-stage game, the MPC is a leader who aims at achieving full cooperation among players by developing a system of incentives at minimum expense. After the leader makes a decision, the followers (different companies) may decide about investment, in cooperative prevention of domino accidents.

1.7.4 Security of Intentional Domino Effects

Reniers et al. [14] proposed to prevent intentional domino effects in chemical clusters in light of possible intentional attacks on chemical infrastructures. Unlike other critical infrastructures, an intentional attack on one or more hazardous installations in chemical industrial areas may trigger escalation effects, resulting in severe consequences. In addition to securing chemical plants, some attempts have been done in the literature. These attempts can be divided into three categories: security of critical installations, mitigation of potential consequences, and reduction of attractiveness.

(1) Security of critical installations

The domino effect potential of hazardous installations depends on the amount of substances present, the physical and toxic properties of the substances, and the specific process conditions, etc. Critical installations refer to the installations with a high likelihood of initiating or propagating accidents. These critical installations are more attractive for adversaries that aim to trigger catastrophic disasters. As a result, the security of critical installations can maximize the benefits of limited protection resources. The strategy of securing critical installations was proposed to address security-related issues in domino effect management [14]. This study indicates that enhancing the security of critical infrastructure can improve the overall security of the chemical industrial area. From the perspective of resilience, Reniers et al. [119] proved that the layout of a chemical industrial area follows a power-law distribution. In other words, only a few installations exhibit very high escalation potential, and securing those installations with high domino effect potential can decrease the possible consequences of an attack. Bubbico and Mazzarotta [120] explored the role of plant layout on security risk in chemical industrial areas, highlighting the role of planning plant layout from a security perspective. According to this study, the most critical zones should be preliminarily identified when planning a plant layout, including process areas, control room(s), storage areas for hazardous materials, loading and unloading facilities, and fire equipment. Based on the concept of the rings of protection, critical facilities should be set in the middle to provide concentric levels of security and increase the number and complexity of the barriers moving toward the center [120].

(2) Mitigation of potential consequences

If the prevention of intentional attacks seems impossible, mitigating the potential consequences by safety measures may be considered an effective approach for protecting chemical industrial plants against domino effects. Safety measures can not only mitigate the potential consequences of intentional attacks but also contribute to the prevention of unintentional domino effects. Therefore, the protection strategy may result in a safer and more secure chemical plant. Based on the protection strategy, Srivastava and Gupta [121] developed a Stepped Matrix Procedure (SMP) method to deal with domino effects by using safety barriers. Reniers and Audenaert [122] proposed to minimize the potential consequences of intentional domino effects based on vulnerability assessment. The approach proposed in the study may also be used to develop an inherent security plant by identifying the most vulnerable installations or to solve layout or site location problems in the early design phase. Khakzad and Reniers [16] proposed a cost-robust mitigation strategy to keep some of the storage tanks empty in the case of imminent terrorist attacks. The robustness of the plant against intentional attacks can be temporarily increased by using this strategy. Consequently, any safety measures for escalation prevention may be used to mitigate the potential consequence of intentional attacks and thus prevent intentional domino effects.

(3) **Reduction of attractiveness**

The attractiveness of hazardous facilities to adversaries depends on many factors: access, security, visibility, opacity, secondary hazard, robustness, law enforcement response, victim profile, and political value [123]. Attractiveness analysis should assess the perceived value of a target to adversaries and the adversaries' choice of targets [124]. Since adversaries usually launch attacks to lead to as many losses as possible [122, 125], both the strategies of security of critical installations and mitigation of potential consequences can lead to a reduction of the attractiveness to adversaries. Therefore, using both safety and security resources to protect chemical industrial areas against intentional domino effects is considered a strategy to reduce attractiveness. Khakzad [125] highlighted the role of reducing the attractiveness of chemical plants to terrorist attacks and recommended using safety concepts such as inherently safer design and land use planning to improve the security of chemical plants. These safety concepts can reduce both the attractiveness of the chemical plant and the consequences of attack scenarios. Chen et al. [15] proposed an integrated approach to prevent and mitigate intentional domino effects, considering the vulnerability of installations exposed to intentional attacks as well as the vulnerability of installations against subsequent domino effects. The developed resource allocation method in this study can effectively reduce a chemical cluster's attractiveness as well as the potential consequences of attacks.

1.8 Research Trends and Future Needs

1.8.1 A Summarization of Current Research

(1) **Risk assessment of domino effects**

Early quantitative research on domino effect assessment mainly forced on the first-level escalations to obtain the likelihood of domino effects in a chemical industrial area [11, 126]. However, only modeling the first-level escalation may underestimate the consequences of escalations while considering second and higher-level escalations is challenging due to the uncertainties associated with higher-level propagations, such as failure types, failure sequences, and intensity of escalation vector. Besides, complex propagations such as synergistic effects and parallel effects may occur at higher-level propagations. In recent years, attempts have been made for modeling higher-level propagations [9, 79]. Graphic methods such as graph metrics and dynamic graphs provide a visible framework for the evolution of domino effects and thus have advantages in modeling higher-level propagation. Graph structures in previous research are mainly based on threshold values, which may only represent one possible evolution sequence of domino effects. Multiple graphs or a more complex structure should be established if the uncertainty of evolution paths is considered. Monte Carlo simulation is a widely used method to deal with propagation uncertainty while the technique may take more computation time than graphic and analytical methods.

Bagster and Pitblado [10] regarded a domino effect accident as a spatial escalation triggered by a loss of containment. Since then, the main task of domino effect risk assessment is modeling the spatial evolution of domino effects, obtaining the likelihood of domino effects and the failure probability of installations [10, 34, 69]. Reniers and Cozzani [6] defined domino effects as a chain of accidents that may occur simultaneously or sequentially, so the evolution of domino effects may be a dynamic process such as the propagation caused by heat radiation (time to failure). Modeling the temporal evolution of domino effects required dynamic tools such as dynamic Bayesian network (DBN) [71], agent-based modeling [80], dynamic graphs [9], and Petri-nets [76]. These dynamic tools were widely used to model superimposed effects in which the heat radiation received by an installation varies in different stages. In terms of domino effects induced by overpressure or fragments, temporal evolution is ignored since the failure caused by these escalation vectors is considered almost instantaneous.

(2) **Domino effect management**

In a chemical cluster, due to possible external escalation effects, the risk of a chemical company depends on the company's safety and security strategies and the decisions of other chemical plants nearby in the chemical cluster. As a result, cooperation management of domino effects is undoubtedly a good choice in a chemical cluster [12, 117]. However, achieving cooperation among different companies is challenging since it is related to technical and organizational problems. To achieve prevention corporation in a chemical cluster, a Multi-Plant Council (MPC) may be recommended to prompt the cooperation [117]. Besides, the decision-making on alert levels in a chemical cluster was also developed to avoid external domino effects [115]. To promote cooperation in a real chemical cluster, more strategic and proactive cooperation should be explored by addressing more organizational issues.

Previous research on domino effect management mainly focused on accidental domino effects using inherent safety [40, 84, 97], safety barriers [99], and emergency response [110, 112, 113]. Little attention has been paid to Natech domino effects and intentional domino effects. Although these protection strategies for accidental domino effects may also be applied to manage domino effects triggered by natural disasters or intentional attacks, special characteristics of these domino effects are not fully addressed. For instance, active protection measures have a high probability of being damaged by natural disasters. Besides, emergency response actions may also be impossible due to the inaccessibility of other critical infrastructures nearby. In terms of intentional domino effects, past management attempts include security of critical installations using security measures [14], mitigation of potential consequences using safety measures [16, 121] and reduction of attractiveness using both safety and security resources [122, 125]. However, adversaries' strategies were almost ignored. For example, multiple-target attacks may lead to multiple fires, resulting in unpreventable escalations.

Based on protection principles, decision-making tools are needed to identify the best strategy according to decision criteria. Decision-making for inherent safety strategies has drawn much attention, including comparative analysis [86, 87], layout optimization [88, 90, 91], and inventory optimization [98]. The decision-making on safety barriers includes optimization of the allocation of safety barriers [104, 105], cost-effective analysis [107] of the influence of costs on decision-making for safety barriers. Besides, game theory is also used to support the decision-making among different plants in a chemical industrial cluster [118]. However, current research mainly focuses on one kind of protection measure or one protection principle, and decision-making based on multiple protection principles which can be used in different stages of chemical plants is lacking. Besides protection costs, the protection benefits may also be interesting for safety and/or security managers since they are important for a company's long-term profitability.

1.8.2 Comparison of Different Modeling Approaches and Protection Strategies

(1) **Comparison of different domino risk assessment approaches**

Many approaches have been proposed for risk assessment and management of domino effects in the process and chemical industry. Different stakeholders related to the process and chemical industry can select different approaches according to their various and different interests and concerns. Consequently, a comparison among different approaches is conducted based on six criteria [1], as follows:

(1) the source of the approach;
(2) the category of the approach;
(3) the vulnerability models used in the approach;
(4) the escalation vector considered in the approach;
(5) the evolution considered in the approach;
(6) the computation cost needed for the implementation of the approach.

According to the analysis in Sect. 1.6, 11 main approaches for modeling domino effects are selected and analyzed based on the six foregoing criteria, as shown in Table 1.9.

(2) **Comparison of different domino management approaches**

Domino effect research aims to prevent and mitigate domino effects in the process and chemical industry. Different protection approaches are available, including inherent safety, safety barriers, emergency response, cooperative prevention in chemical industrial clusters, and security of domino effects. To prevent domino effects, the possible causes of domino effects should be analyzed since different areas may face

Table 1.9 Comparison among different modeling approaches (Chen et al. [1])

Source	Category	Vulnerability basis	Escalation vector	Evolution	Computation cost
Bagster and Pitblado [10]	Analytical method	Safety distances	Multiple	Higher-level	High in large scale case
Khan and Abbasi [34]	Analytical method	Probabilistic models	Multiple	First-level	Low
Cozzani et al. [11, 17]	Analytical method	Probit models	Multiple	First-level Extend to higher levels	Low High in large scale case
Reniers and Dullaert [13]	Network method	Safety distances	Multiple	Higher-level	low
Abdolhamidzadeh et al. [79, 126]	Model-Carlo simulation	Probit models	Multiple	First-level Extend to higher levels	high
Khakzad et al. [69]	Bayesian network	Probit models and thresholds	Multiple	Higher-level	High in large scale case
Khakzad [71]	Dynamic Bayesian network	Probit models and thresholds	Heat radiation	Higher-level Temporal evolution	High in large scale case
Khakzad and Reniers [66]	Graph metrics	thresholds	Multiple	Higher-level	Low
Kamil et al. [76] Zhou and Reniers [75]	Petri-net	Probit models	Multiple	Higher-level	Low High
Zhang et al. [80]	Agent-based simulation	Probit models	Heat radiation	Higher-level Temporal evolution	High
Chen et al. [15]	Dynamic graph model	Residual time to failure	Heat radiation	Higher-level Temporal evolution	Low

different threats. For accidental domino effects, hazards are mainly located within a chemical industrial area, while in the case of intentional domino effects, from outside the chemical industrial park threats may be involved. Safety measures for escalation prevention (i.e., inherent safety, safety barriers, and emergency response) can reduce the risk of domino effects caused by intentional attacks, accidental events, and natural disasters. Security measures are mainly for preventing intentional attacks and thus reduce the probability of intentional domino effects. Besides, cooperative prevention strategies are proposed to prevent possible external domino effects, enhancing the safety and security of chemical industrial clusters. Therefore, stakeholders should choose protection strategies according to their threats, concerns, and preferences, as shown in Fig. 1.7.

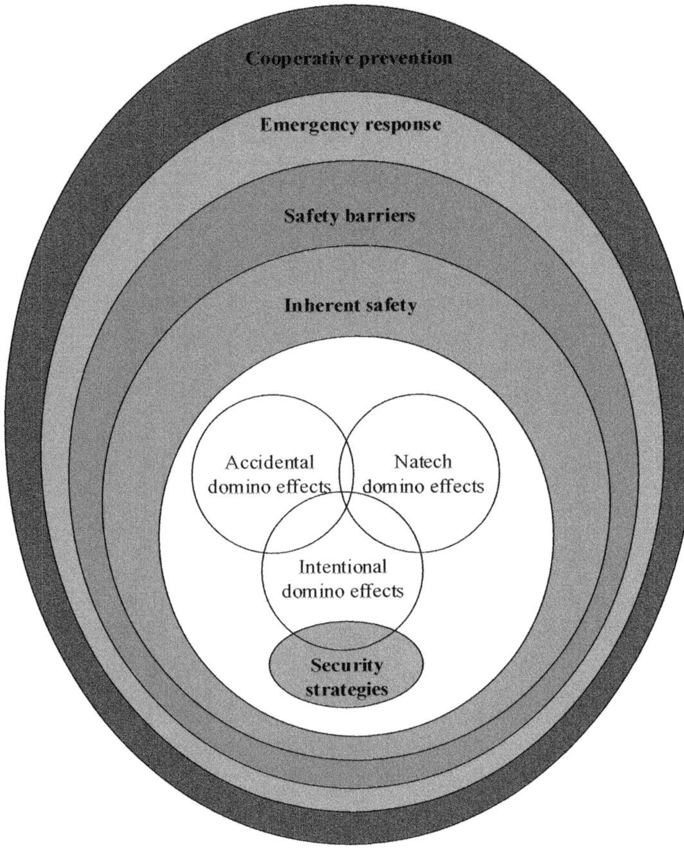

Fig. 1.7 Protection strategies for preventing and mitigating domino effects

1.8.3 Research Gaps

According to the review of past work on risk assessment and management of domino effects in the process and chemical industry, research gaps are identified.

(1) **Dynamic risk assessment of fire-induced domino effects**

The propagation of fire scenarios depends on the time to failure (ttf) of installations exposed to fire. As a result, fire-induced escalation may be regarded as a spatial–temporal evolution process. During the evolution, one installation may receive heat radiation from multiple fires (synergistic effects), and the received heat radiation may change over time. The effects of heat radiation in different stages should be superimposed when determining the ttf (superimposed effects). Besides, the time-lapse in the second or higher-level escalation should be considered in probit models. In light of these research gaps, Chap. 2 develops a dynamic graph approach to model fire-induced domino effects.

(2) **Dynamic risk assessment of VCE-induced domino effects**

Compared with fire scenarios, vapor cloud explosion (VCE) is more difficult to assess due to the uncertainty of ignition position, the uncertainty of delayed ignition time (DIT), and the complexity of overpressure intensity calculation. The VCE induced by the release of hazardous substances in chemical plants is a dynamic process along with the vapor cloud dispersion. However, previous risk analysis methods for VCE always assume that the explosion occurs immediately at the release place [75, 126], which is inconsistent with the observations from large VCEs in recent years. As a result, a dynamic tool is established in Chap. 3 to address the vapor cloud dispersion and delayed ignition in the assessment of VCE-induced domino effects.

(3) **Dynamic risk assessment of coupled domino effects**

If a loss of containment occurs in a chemical industrial area, accident scenarios such as a toxic release, a VCE, and a fire may simultaneously or sequentially occur, and the generated hazards can evolve spatially and result in a cascading disaster. Consequently, all the major hazards (fire, explosion, and toxic release) can simultaneously or sequentially be present in a domino effect. Neglecting any known hazard may underestimate the risk of domino effects and result in more severe consequences. Therefore, modeling the spatial–temporal evolution of hazards originating from the release of hazardous materials in industrial areas is essential for protecting staff, nearby residents, and emergency rescuers. As a result, a dynamic method is developed in Chap. 4 to model coupled domino effects.

(4) **Integrated management of domino effects**

Domino effects can be triggered by intentional or unintentional events. Safety barriers can reduce the likelihood and consequences of accidental domino effects, Natech domino effects, and intentional domino effects. Security resources are essential to prevent intentional attacks and also decrease the attractiveness of a possible target. Past research on domino effect management mainly concerned unintentional domino effects, neglecting intentional domino effects, which may result in even more severe consequences. Thus safety and security resources may be integrated to manage different kinds of domino effects from a systemic perspective. Therefore, Chap. 5 establishes an integrated management framework to prevent and mitigate domino effects.

(5) **Economic approach to manage domino effects**

Many safety and security measures can be used for managing domino effects, while not all of these measures can be implemented due to the safety and security budget. Chemical companies have to consider the costs of protection measures since the budget is not infinite, evidentially. Consequently, the economic issues of safety and security play an indispensable role in the decision-making on the allocation of safety and security measures. In decision-making on the investment in the prevention and mitigation of domino effects, economic approaches may be used to address these economic issues and make the protection more profitable. Chapter 6 thus introduces an economic approach to manage domino effects, considering both the costs of protection measures and domino effect events.

(6) **Resilience approach to manage domino effects**

Disruptions that may trigger domino effects such as intentional attacks may be difficult to predict and prevent, thus safety and security measures may be insufficient for preventing domino effects. Once a domino effect occurs, an adaptation operation or a quick restoration can reduce the loss and thus mitigate the consequences of domino effects. Resilience refers to the capability of a chemical plant to resist, mitigate, adapt, and recover from undesired events, to maintain its desired performance. As a result, developing a resilient chemical plant may be a practical and effective way to deal with these disruptions. A resilience-based approach is proposed in Chap. 7 to prepare a chemical plant to anticipate, absorb, adapt to, and restore from domino accidents.

1.9 Conclusions

In recent decades, the importance of domino effect management has been recognized by researchers and practitioners in the process and chemical industries. More and more efforts have been put into assessing the possibility of domino effects, modeling the evolution of domino effects, and preventing or mitigating domino effects. This chapter reviews the research issues and methods in domino effect risk assessment and management and their development in the literature. The available methods for simulating the domino effect are roughly divided into three categories: analytical methods, graphical methods, and simulation methods. The current management strategies are divided into five types: inherent security, security barrier management, emergency response, cooperative prevention, and security strategies. It provides a very clear picture of the development of research issues and the methods used to assess and manage possible domino effects. Besides, according to several standards, different types of modeling methods and management strategies are further compared to locate their applications and promote future research directions. Finally, we obtain research gaps, which are also the motivations for the following chapters.

References

1. Chen C, Reniers G, Khakzad N (2020) A thorough classification and discussion of approaches for modeling and managing domino effects in the process industries. Saf Sci 125
2. Chen C (2021) A dynamic and integrated approach for modeling and managing domino-effects. Delft University of Technology

3. Mukhim ED, Abbasi T, Tauseef SM, Abbasi SA (2017) Domino effect in chemical process industries triggered by overpressure—formulation of equipment-specific probits. Process Saf Environ Prot 106:263–273
4. Necci A, Cozzani V, Spadoni G, Khan F (2015) Assessment of domino effect: state of the art and research Needs. Reliab Eng Syst Saf 143:3–18
5. Reniers GL, Van Erp HN (2016) Operational safety economics: a practical approach focused on the chemical and process industries. Wiley
6. Reniers G, Cozzani V (2013) Domino effects in the process industries, modeling, prevention and managing. Elsevier, Amsterdam, The Netherlands
7. Arturson GJB (1987) The tragedy of San Juanico—the most severe LPG disaster in history. 13(2):87–102
8. Chen C, Khakzad N, Reniers G (2020) Dynamic vulnerability assessment of process plants with respect to vapor cloud explosions. Reliab Eng Syst Saf 200
9. Chen C, Reniers G, Zhang L (2018) An innovative methodology for quickly modeling the spatial-temporal evolution of domino accidents triggered by fire. J Loss Prev Process Ind 54:312–324
10. Bagster DF, Pitblado RM (1991) The estimation of domino incident frequencies—an approach. Process Saf Environ Prot 69(4):195–199
11. Cozzani V, Gubinelli G, Antonioni G, Spadoni G, Zanelli S (2005) The assessment of risk caused by domino effect in quantitative area risk analysis. J Hazard Mater 127(1–3):14–30
12. Reniers GLL, Dullaert W, Ale BJM, Soudan K (2005) Developing an external domino accident prevention framework: Hazwim. J Loss Prev Process Ind 18(3):127–138
13. Reniers GLL, Dullaert W (2007) DomPrevPlanning©: user-friendly software for planning domino effects prevention. Saf Sci 45(10):1060–1081
14. Reniers G, Dullaert W, Audenaert A, Ale BJ, Soudan K (2008) Managing domino effect-related security of industrial areas. J Loss Prev Process Ind 21(3):336–343
15. Chen C, Reniers G, Khakzad N (2019) Integrating safety and security resources to protect chemical industrial parks from man-made domino effects: a dynamic graph approach. Reliab Eng Syst Saf 191
16. Khakzad N, Reniers G (2019) Low-capacity utilization of process plants: a cost-robust approach to tackle man-made domino effects. Reliab Eng Syst Saf 191
17. Cozzani V, Antonioni G, Landucci G, Tugnoli A, Bonvicini S, Spadoni G (2014) Quantitative assessment of domino and NaTech scenarios in complex industrial areas. J Loss Prev Process Ind 28:10–22
18. Yang Y, Chen G, Chen P (2018) The probability prediction method of domino effect triggered by lightning in chemical tank farm. Process Saf Environ Prot 116:106–114
19. Papadakis G, Amendola A (1997) Guidance on the preparation of a safety report to meet the requirements of Council Directive 96/82/EC (Seveso II). Joint Research Centre, European Commission
20. Council Directive 2012/18/EU (2012) On the control of major-accident hazards involving dangerous substances. Off J Eur Commun 1–38
21. Darbra RM, Palacios A, Casal J (2010) Domino effect in chemical accidents: main features and accident sequences. J Hazard Mater 183(1–3):565–573
22. Hemmatian B, Abdolhamidzadeh B, Darbra RM, Casal J (2014) The significance of domino effect in chemical accidents. J Loss Prev Process Ind 29:30–38
23. Abdolhamidzadeh B, Abbasi T, Rashtchian D, Abbasi SA (2011) Domino effect in process-industry accidents—an inventory of past events and identification of some patterns. J Loss Prev Process Ind 24(5):575–593
24. Alileche N, Cozzani V, Reniers G, Estel L (2015) Thresholds for domino effects and safety distances in the process industry: a review of approaches and regulations. Reliab Eng Syst Saf 143:74–84
25. Li J, Reniers G, Cozzani V, Khan F (2017) A bibliometric analysis of peer-reviewed publications on domino effects in the process industry. J Loss Prev Process Ind 49:103–110

26. Swuste P, van Nunen K, Reniers G, Khakzad N (2019) Domino effects in chemical factories and clusters: an historical perspective and discussion. Process Saf Environ Prot
27. Jones ML (2004) Application of systematic review methods to qualitative research: practical issues. J Adv Nurs 48(3):271–278
28. Lees FP (1996) Loss prevention in the process industries. Butterworth-Heinemann, Oxford
29. Council Directive 96/82/EC (1997) Council Directive 96/82/EC of 9 December 1996 on the control of major-accident hazards involving dangerous substances. Off J Eur Commun 1–38
30. Khan FI, Abbasi SA (1998) DOMIFFECT (DOMIno eFFECT): user-friendly software for domino effect analysis. Environ Model Softw 13(2):163–177
31. Reniers GLL, Dullaert W, Ale BJM, Soudan K (2005) The use of current risk analysis tools evaluated towards preventing external domino accidents. J Loss Prev Process Ind 18(3):119–126
32. Salzano E, Cozzani V (2003) The use of probit functions in the quantitative risk assessment of domino accidents caused by overpressure. Saf Reliab 1 and 2. A Balkema Publishers, Leiden
33. Gomez-Mares M, Zarate L, Casal J (2008) Jet fires and the domino effect. Fire Saf J 43(8):583–588
34. Khan FI, Abbasi S (1998) Models for domino effect analysis in chemical process industries. Process Saf Prog 17(2):107–123
35. Reniers G, Khakzad N, Cozzani V, Khan F (2018) The impact of nature on chemical industrial facilities: dealing with challenges for creating resilient chemical industrial parks. J Loss Prev Process Ind
36. Reniers G (2010) An external domino effects investment approach to improve cross-plant safety within chemical clusters. J Hazard Mater 177(1–3):167–174
37. Moodie K (1988) Experiments and modelling: an overview with particular reference to fire engulfment. J Hazard Mater 20:149–175
38. Uijt de Haag, Ale (1999) Guidelines for quantitative risk assessment. Committee for the Prevention of Disasters, The Hague (NL)
39. Van Den Bosh C, Weterings R (1997) Methods for the calculation of physical effects (Yellow Book). Committee for the Prevention of Disasters, The Hague (NL)
40. Cozzani V, Tugnoli A, Salzano E (2009) The development of an inherent safety approach to the prevention of domino accidents. Accid Anal Prev 41(6):1216–1227
41. Lees F (2012) Lees' Loss prevention in the process industries: hazard identification, assessment and control. Butterworth-Heinemann
42. Cozzani V, Salzano E (2004) The quantitative assessment of domino effects caused by overpressure: Part I. Probit models. J Hazard Mater 107(3):67–80
43. Zhang M, Jiang J (2008) An improved probit method for assessment of domino effect to chemical process equipment caused by overpressure. J Hazard Mater 158(2–3):280–286
44. Landucci G, Reniers G, Cozzani V, Salzano E (2015) Vulnerability of industrial facilities to attacks with improvised explosive devices aimed attriggering domino scenarios. Reliab Eng Syst Saf 143:53–62
45. Salzano E, Cozzani V (2006) A fuzzy set analysis to estimate loss intensity following blast wave interaction with process equipment. J Loss Prev Process Ind 19(4):343–352
46. Tugnoli A, Gubinelli G, Landucci G, Cozzani V (2014) Assessment of fragment projection hazard: probability distributions for the initial direction of fragments. J Hazard Mater 279:418–427
47. Gubinelli G, Zanelli S, Cozzani V (2004) A simplified model for the assessment of the impact probability of fragments. J Hazard Mater 116(3):175–187
48. Hauptmanns U (2001) A Monte-Carlo based procedure for treating the flight of missiles from tank explosions. Probab Eng Mech 16(4):307–312
49. Zhang X-m, Chen G-h (2009) The analysis of domino effect impact probability triggered by fragments. Saf Sci 47(7):1026–1032
50. Landucci G, Gubinelli G, Antonioni G, Cozzani V (2009) The assessment of the damage probability of storage tanks in domino events triggered by fire. Accid Anal Prev 41(6):1206–1215

51. Rum A, Landucci G, Galletti C (2018) Coupling of integral methods and CFD for modeling complex industrial accidents. J Loss Prev Process Ind 53:115–128
52. Landucci G, Molag M, Cozzani V (2009) Modeling the performance of coated LPG tanks engulfed in fires. J Hazard Mater 172(1):447–456
53. Jujuly M, Rahman A, Ahmed S, Khan F (2015) LNG pool fire simulation for domino effect analysis. Reliab Eng Syst Saf 143:19–29
54. Dasgotra A, Varun Teja GVV, Sharma A, Mishra KB (2018) CFD modeling of large-scale flammable cloud dispersion using FLACS. J Loss Prev Process Ind 56:531–536
55. Ahmadi O, Mortazavi SB, Pasdarshahri H, Mohabadi HA (2019) Consequence analysis of large-scale pool fire in oil storage terminal based on computational fluid dynamic (CFD). Process Saf Environ Prot 123:379–389
56. Assael MJ, Kakosimos KE (2010) Fires, explosions, and toxic gas dispersions: effects calculation and risk analysis. CRC Press
57. Khan F, Abbasi S (1996) Simulation of accidents in a chemical industry using the software package MAXCRED. Indian J Chem Technol 3(6):338–344
58. Khan FI, Iqbal A, Abbasi SA (2001) Risk analysis of a petrochemical industry using ORA (Optimal Risk Analysis) procedure. Process Saf Prog 20(2):95–110
59. Cozzani V, Antonioni G, Spadoni G (2006) Quantitative assessment of domino scenarios by a GIS-based software tool. J Loss Prev Process Ind 19(5):463–477
60. van der Voort MM, Klein AJJ, de Maaijer M, van den Berg AC, van Deursen JR, Versloot NHA (2007) A quantitative risk assessment tool for the external safety of industrial plants with a dust explosion hazard. J Loss Prev Process Ind 20(4–6):375–386
61. Zhang X-m, Chen C (2013) Mechanism analysis and risk assessment of escalation scenario in chemical industry zones. Process Saf Environ Prot 91(1–2):79–85
62. Kadri F, Châtelet E, Chen G (2013) Method for quantitative assessment of the domino effect in industrial sites. Process Saf Environ Prot 91(6):452–462
63. Zhou J, Reniers G (2018) A matrix-based modeling and analysis approach for fire-induced domino effects. Process Saf Environ Prot 116:347–353
64. Ramirez-Camacho JG, Pastor E, Casal J, Amaya-Gomez R, Munoz-Giraldo F (2015) Analysis of domino effect in pipelines. J Hazard Mater 298:210–220
65. Silva EP, Nele M, e Melo PFF, Könözsy L (2016) Underground parallel pipelines domino effect: an analysis based on pipeline crater models and historical accidents. J Loss Prev Process Ind 43:315–331
66. Khakzad N, Reniers G (2015) Using graph theory to analyze the vulnerability of process plants in the context of cascading effects. Reliab Eng Syst Saf 143:63–73
67. Li, Wang, Ge, Wei, Li (2019) Risk analysis of earth-rock dam breach based on dynamic Bayesian network. Water 11(11)
68. Chen P, Huang Y, Mou J, van Gelder PHAJM (2019) Probabilistic risk analysis for ship-ship collision: state-of-the-art. Saf Sci 117:108–122
69. Khakzad N, Khan F, Amyotte P, Cozzani V (2013) Domino effect analysis using Bayesian networks. Risk Anal 33(2):292–306
70. Khakzad N, Amyotte P, Cozzani V, Reniers G, Pasman H (2018) How to address model uncertainty in the escalation of domino effects? J Loss Prev Process Ind 54:49–56
71. Khakzad N (2015) Application of dynamic Bayesian network to risk analysis of domino effects in chemical infrastructures. Reliab Eng Syst Saf 138:263–272
72. Khakzad N, Dadashzadeh M, Reniers G (2018) Quantitative assessment of wildfire risk in oil facilities. J Environ Manag 223:433–443
73. Chen C, Reniers G, Khakzad N (2021) A dynamic multi-agent approach for modeling the evolution of multi-hazard accident scenarios in chemical plants. Reliab Eng Syst Saf 207
74. David R, Alla H (2005) Discrete, continuous, and hybrid Petri nets, vol 1. Springer
75. Zhou J, Reniers G (2017) Petri-net based cascading effect analysis of vapor cloud explosions. J Loss Prev Process Ind 48:118–125
76. Kamil MZ, Taleb-Berrouane M, Khan F, Ahmed S (2019) Dynamic domino effect risk assessment using Petri-nets. Process Saf Environ Prot 124:308–316

77. Ferdous R, Khan F, Sadiq R, Amyotte P, Veitch B (2009) Handling data uncertainties in event tree analysis. Process Saf Environ Prot 87(5):283–292
78. Alileche N, Olivier D, Estel L, Cozzani V (2017) Analysis of domino effect in the process industry using the event tree method. Saf Sci 97:10–19
79. Rad A, Abdolhamidzadeh B, Abbasi T, Rashtchian D (2014) FREEDOM II: an improved methodology to assess domino effect frequency using simulation techniques. Process Saf Environ Prot 92(6):714–722
80. Zhang L, Landucci G, Reniers G, Khakzad N, Zhou J (2018) DAMS: a model to assess domino effects by using agent-based modeling and simulation. Risk Anal 38(8):1585–1600
81. Bollinger RE, Crowl DA (1997) Inherently safer chemical processes: a life cycle approach. Wiley-AIChE
82. Khan FI, Amyotte PR (2003) How to make inherent safety practice a reality. Can J Chem Eng 81(1):2–16
83. Kletz TA (2003) Inherently safer design—its scope and future. Process Saf Environ Prot 81(6):401–405
84. Cozzani V, Tugnoli A, Salzano E (2007) Prevention of domino effect: from active and passive strategies to inherently safer design. J Hazard Mater 139(2):209–219
85. Tugnoli A, Khan F, Amyotte P, Cozzani V (2008) Safety assessment in plant layout design using indexing approach: implementing inherent safety perspective. Part 1—Guideword applicability and method description. J Hazard Mater 160(1):100–109
86. Tugnoli A, Khan F, Amyotte P, Cozzani V (2008) Safety assessment in plant layout design using indexing approach: implementing inherent safety perspective. Part 2-Domino hazard index and case study. J Hazard Mater 160(1):110–121
87. Landucci G, Tugnoli A, Cozzani V (2008) Inherent safety key performance indicators for hydrogen storage systems. J Hazard Mater 159(2–3):554–566
88. Lee JY, Lee JW, Ko JW, Yoon ES (2005) Optimization for allocating the explosive facilities in order to minimize optimize the domino effect using nonlinear programming. Korean J Chem Eng 22(5):649–656
89. Lee JY, Kim HS, Yoon ES (2006) A new approach for allocating explosive facilities in order to minimize the domino effect using NLP. J Chem Eng Jpn 39(7):731–745
90. So W, Kim Y-H, Lee CJ, Shin D, Yoon ES (2011) Optimal layout of additional facilities for minimization of domino effects based on worst-case scenarios. Korean J Chem Eng 28(3):656–666
91. Dan S, Shin D, Yoon ES (2015) Layout optimization of LNG-liquefaction process on LNG-FPSO preventing domino effects. J Chem Eng Jpn 48(8):646–655
92. Jung S, Ng D, Diaz-Ovalle C, Vazquez-Roman R, Mannan MS (2011) New approach to optimizing the facility siting and layout for fire and explosion scenarios. Ind Eng Chem Res 50(7):3928–3937
93. de Lira-Flores J, Vázquez-Román R, López-Molina A, Mannan MS (2014) A MINLP approach for layout designs based on the domino hazard index. J Loss Prev Process Ind 30:219–227
94. Latifi SE, Mohammadi E, Khakzad N (2017) Process plant layout optimization with uncertainty and considering risk. Comput Chem Eng 106:224–242
95. Bernechea EJ, Arnaldos J (2014) Optimizing the design of storage facilities through the application of ISD and QRA. Process Saf Environ Prot 92(6):598–615
96. Nomen R, Sempere J, Mariotti V (2014) QRA including domino effect as a tool for engineering design. Procedia Engineering 84:23–32
97. Khakzad N, Reniers G (2015) Risk-based design of process plants with regard to domino effects and land use planning. J Hazard Mater 299:289–297
98. Khakzad N, Khan F, Amyotte P, Cozzani V (2014) Risk management of domino effects considering dynamic consequence analysis. Risk Anal 34(6):1128–1138
99. Landucci G, Argenti F, Tugnoli A, Cozzani V (2015) Quantitative assessment of safety barrier performance in the prevention of domino scenarios triggered by fire. Reliab Eng Syst Saf 143:30–43

100. Landucci G, Argenti F, Spadoni G, Cozzani V (2016) Domino effect frequency assessment: the role of safety barriers. J Loss Prev Process Ind 44:706–717
101. Landucci G, Bonvicini S, Cozzani V (2017) A methodology for the analysis of domino and cascading events in oil & gas facilities operating in harsh environments. Saf Sci 95:182–197
102. Khan F, Ahmed S, Yang M, Hashemi SJ, Caines S, Rathnayaka S, Oldford D (2015) Safety challenges in harsh environments: lessons learned. Process Saf Prog 34(2):191–195
103. Khakzad N, Landucci G, Reniers G (2017) Application of dynamic Bayesian network to performance assessment of fire protection systems during domino effects. Reliab Eng Syst Saf 167:232–247
104. Tugnoli A, Cozzani V, Di Padova A, Barbaresi T, Tallone F (2012) Mitigation of fire damage and escalation by fireproofing: a risk-based strategy. Reliab Eng Syst Saf 105:25–35
105. Ghasemi AM, Nourai F (2017) A framework for minimizing domino effect through optimum spacing of storage tanks to serve in land use planning risk assessments. Saf Sci 97:20–26
106. Janssens J, Talarico L, Reniers G, Sörensen K (2015) A decision model to allocate protective safety barriers and mitigate domino effects. Reliab Eng Syst Saf 143:44–52
107. Khakzad N, Landucci G, Cozzani V, Reniers G, Pasman H (2018) Cost-effective fire protection of chemical plants against domino effects. Reliab Eng Syst Saf 169:412–421
108. Cincotta S, Khakzad N, Cozzani V, Reniers G (2019) Resilience-based optimal firefighting to prevent domino effects in process plants. J Loss Prev Process Ind 58:82–89
109. Zhou J, Reniers G (2016) Petri-net based simulation analysis for emergency response to multiple simultaneous large-scale fires. J Loss Prev Process Ind 40:554–562
110. Zhou J, Reniers G (2018) Petri-net based evaluation of emergency response actions for preventing domino effects triggered by fire. J Loss Prev Process Ind 51:94–101
111. Zhou J, Reniers G (2017) Analysis of emergency response actions for preventing fire-induced domino effects based on an approach of reversed fuzzy Petri-net. J Loss Prev Process Ind 47:169–173
112. Zhou J, Reniers G, Khakzad N (2016) Application of event sequence diagram to evaluate emergency response actions during fire-induced domino effects. Reliab Eng Syst Saf 150:202–209
113. Khakzad N (2018) Which fire to extinguish first? A risk-informed approach to emergency response in oil terminals. Risk Anal 38(7):1444–1454
114. Khakzad N (2018) A graph theoretic approach to optimal firefighting in oil terminals. Energies 11(11)
115. Hosseinnia B, Khakzad N, Reniers G (2018) Multi-plant emergency response for tackling major accidents in chemical industrial areas. Saf Sci 102:275–289
116. Reniers G, Soudan K (2010) A game-theoretical approach for reciprocal security-related prevention investment decisions. Reliab Eng Syst Saf 95(1):1–9
117. Reniers G, Dullaert W, Karel S (2009) Domino effects within a chemical cluster: a game-theoretical modeling approach by using Nash-equilibrium. J Hazard Mater 167(1–3):289–293
118. Pavlova Y, Reniers G (2011) A sequential-move game for enhancing safety and security cooperation within chemical clusters. J Hazard Mater 186(1):401–406
119. Reniers GLL, Sörensen K, Khan F, Amyotte P (2014) Resilience of chemical industrial areas through attenuation-based security. Reliab Eng Syst Saf 131:94–101
120. Bubbico R, Mazzarotta B (2014) Security risk assessment of process plants: the role of layout. In: Garzia F, Thomas G, Pritchard DA (eds) 2014 International Carnahan conference on security technology. International Carnahan conference on security technology proceedings. IEEE, New York
121. Srivastava A, Gupta JP (2010) New methodologies for security risk assessment of oil and gas industry. Process Saf Environ Prot 88(6):407–412
122. Reniers GLL, Audenaert A (2014) Preparing for major terrorist attacks against chemical clusters: Intelligently planning protection measures w.r.t. domino effects. Process Saf Environ Prot 92(6):583–589
123. Coster MN, Hankin RK (2003) Risk assessment of antagonistic hazards. J Loss Prev Process Ind 16(6):545–550

124. API (2013) ANSI/API Standard 780—Security risk assessment methodology for the petroleum and petrochemical industry. American Petroleum Institute
125. Khakzad N (2018) Reducing the attractiveness of chemical plants to terrorist attacks: dehorning rhinos. Process Saf Prog 37(2):150–152
126. Abdolhamidzadeh B, Abbasi T, Rashtchian D, Abbasi SA (2010) A new method for assessing domino effect in chemical process industry. J Hazard Mater 182(1–3):416–426

Chapter 2
Dynamic Risk Assessment of Fire-Induced Domino Effects

2.1 Introduction

In chemical industrial areas,[1] a loss of containment (LOC) of flammable materials can lead to a fire if ignited immediately [4]. Fire is the most common primary scenario of domino effects, accounting for 52.4% of the escalation events in process and storage plants [5]. As a result, many studies have been done to assess fire-induced domino effects, and many safety criteria are based on fire escalation [6]. The International Oil and Gas Producers Association (OGP) provides an escalation threshold of 25 kW/m^2 for the failure of equipment with unprotected steel and 35 kW/m^2 for equipment with cellulosic materials [7]. These thresholds are very simple and user-friendly and thus widely adopted in the engineering domain. However, an assumption that fire propagation is not time-dependent is hidden in threshold-based methods, which may underestimate the likelihood of fire-induced domino effects.

The failure of installations exposed to heat radiation needs to suffer a heat-up process, and the time-lapse between the beginning of exposure and the failure is called time to failure (ttf) [8]. Consequently, the vulnerability of installations exposed to fire depends on both the intensity of heat radiation and the exposure time. For a particular vessel, the ttf decreases with increasing the intensity of heat radiation. Landucci et al. [8] established equations for calculating the ttf of vessels under different intensities of heat radiation. According to the ttf calculation equations and considering the effects of emergency response on preventing fire propagation, probit models are developed to assess the fire-level propagation caused by fire. The probit models are established based on the following assumption: an escalation probability of 0.1 is assumed to be comparable to a ttf equal to 5 min. That of 0.9 is assigned to a ttf equal to 20 min. Since the heat radiation received by installations may change over time and a time-lapse exists between first-level escalation and higher-level escalation, the "probit model" may be unreasonable for accurately estimating the damage probability of

[1] This chapter is mainly based on two publications: Chen et al. [1, 2] and Chen [3].

C. Chen et al., *Integrating Safety and Security Management to Protect Chemical Industrial Areas from Domino Effects*, Springer Series in Reliability Engineering,
https://doi.org/10.1007/978-3-030-88911-1_2

installations in second-level or higher-level propagations. In other words, using the model may result in an overestimation of the probability propagation in secondary or higher levels.

The cause of the primary fire may be accidental events or intentional attacks. The former is related to safety, while the latter involves security. Compared with accidental events, intentional events may be more likely to lead to multiple primary fires due to possible simultaneous attacks on multiple installations. In terms of integrating safety and security, the risk assessment model for fire-induced domino effects should therefore address multiple primary fires.

In a chemical industrial area, various hazardous installations are located nearby with different domino effect potentials. Some installations may have a high probability of initiating domino accidents, while other installations are likely to propagate domino events. Application of graph/ network theory, these installations can be modeled as nodes. The quantitative possibility of accident propagation may be represented by the weight of the edges between nodes in a network or graph [9]. Based on this concept, critical installations contributing to possible domino effects can be identified, and this information may support the allocation of domino prevention resources [10, 11]. This concept can be extended for vulnerability analysis and protection decision-making using graph metrics (e.g., betweenness and closeness) [12–14]. These graph-based models can assess domino risks within an entire industrial area and quickly identify the most critical units. However, they fail to capture the time dependencies and uncertainties in fire escalations. Besides, Bayesian network was also applied to model domino effect propagation [15, 16]. The Bayesian network model can tackle higher-level propagations while it may be difficult to apply it to chemical clusters with a large number of installations [12]. Furthermore, simulation-based approaches were proposed based on the Monte Carlo method to model higher-level propagations. The approach successfully models the spatial evolution of domino accidents, but the shortcoming is obvious, i.e., it is time-consuming [17–20].

In light of the limitations of previous work and the need for integrating safety and security, a dynamic graph approach [1, 2] is introduced in this chapter to model fire escalations, assessing the risk of fire-induced domino effects. Section 2.2 introduces the characteristics of fire escalation, and the theory of dynamic graphs is presented in Sect. 2.3. A domino evolution graph (DEG) model based on dynamic graphs for modeling domino effects triggered by fire is illustrated in Sect. 2.4. Conclusions are drawn in Sect. 2.8.

2.2 Fire Escalation

The majority of escalation events in the chemical and process industry are triggered by fire events such as pool fires and jet fires [21]. These primary scenarios can cause significant heat loads on installations, potentially leading to the failure of process units, storage vessels, pipework, and other target installations through one or more escalation vectors: heat radiation, fire engulfment, and fire impingement [7, 8]. Heat flux can permeate a vessel wall and lead to the build-up of temperature and pressure within the vessel exposed to the fire. The temperature and pressure of the vessel increase over time and may lead to a loss of containment when the vessel fails. Usually, escalation caused by fire occurs when a loss of containment emerges. If the released hazardous materials are ignited immediately, the damaged tank catches fire and may cause more heat flux on other undamaged vessels nearby. The fire spatially escalates over time. As a result, fire escalation is a spatial–temporal evolution of fire.

2.2.1 *Spatial Escalation*

Escalation is the main characteristic of domino accidents, linking a primary scenario with one or more than one higher-order scenarios [19]. Based on the spatial escalation characteristics, escalation patterns can be divided into three categories: simple escalation, multilevel domino chain, and multilevel escalation [22], as shown in Fig. 2.1.

(1) Simple escalation: a simple escalation is the domino effect that only a primary scenario triggers a secondary scenario (a "one-to-one" pattern). In the simple escalation, no higher escalations exist, and only two installations are damaged.

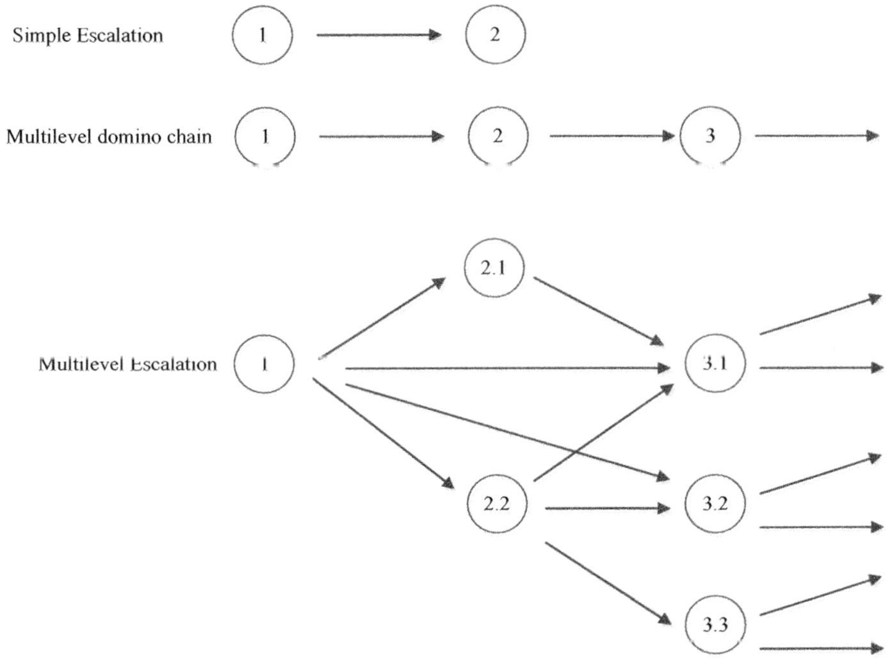

Fig. 2.1 Example of escalation patterns (Chen et al. [1])

(2) Multilevel chain escalation: a multilevel chain escalation consists of a chain of accidents in which a primary scenario triggers a second accident scenario, and the second accident scenario triggers a third accident scenario, and so on. In terms of this pattern, protecting one of the installations in the chain can successfully prevent higher-level escalations.

(3) Multilevel escalation: a multilevel escalation consists of multiple chains. This pattern is a complex pattern in which multiple accident scenarios may trigger one higher-order scenario while multiple accident scenarios can also be induced by one lower-order accident scenario. The former escalation is called the "synergistic effect" while the latter escalation is called the "parallel effect". Due to the two complex effects, modeling multilevel escalations is challenging. As shown in Fig. 2.1, fire at installation 3.1 is triggered by fires at installations 1, 2.1, and 2.2, which is a synergistic effect. The fire at installation 1 induces fires at installations 2.1 and 2.2, which is a parallel effect. The parallel effect shows the escalation capability of an installation to initial and propagate domino effects, while the synergistic effect reflects the vulnerability of an installation. Therefore, the parallel effect and the synergistic effect should be addressed in modeling the spatial escalation of domino effects.

2.2.2 Temporal Escalation

Since there is a time-lapse between the start of exposure and the failure caused by heat radiation, the escalation caused by fire is a time-dependent process. Using a simple escalation (a hazardous material vessel to a hazardous material vessel) as an example, the fire escalation process can be explained as follows:

- The heat caused by a fire is transferred to the vessel wall (shell) by heat radiation or heat convection;
- The target vessel wall is heated, and the temperature rises with the heating;
- The heat transfers from the vessel wall to the liquid and liquid–vapor due to the existing temperature gradients among the vessel wall, the liquid, and the liquid–vapor inside the vessel;
- The internal pressure and temperature of the target vessel increase with the build-up of heat;
- With the heating, the target vessel may fail, resulting in a loss of containment of hazardous materials.

The escalation caused by fire occurs when the target vessel fails. Due to the build-up of heat, the escalation is delayed compared to the time that the target vessel starts to exposure to fire. The time-lapse between the start of the fire and the failure of the target vessel is called "time to failure" (*ttf*) [6]. The burning time of a vessel is called "time to burn out" (*ttb*) [23]. If $ttf < ttb$, then fire escalation may occur; otherwise, the heating ends and the escalation does not occur. Besides the *ttf* and *ttb*, the escalation caused by heat radiation also depends on safety barriers such as emergency response actions. There is a time-lapse between the start of a fire and the time to control the fire using emergency response. The time-lapse is called "time to control" (*ttc*). Since the escalation of fire depends on multiple temporal factors, it is also a temporal propagation.

According to the characteristics of the escalation induced by fire, a fire-induced domino effect is a spatial–temporal evolution process of fire scenarios in a chemical industrial area. To model fire-induced domino effects, we need to integrate spatial escalation and temporal escalation using a dynamic approach. As a result, a dynamic approach based on graph theory is developed to assess fire-induced domino effects, addressing the spatial–temporal evolution process.

2.3 Graph Theory

2.3.1 Static Graphs

Graph theory is a mathematical method for studying the interconnections between elements in natural and artificial systems. Initially, entity interactions were restricted to binary relationships denoted by graph vertices. Functions were then correlated with graphs, assigning a real number to each graph edge to measure the relationship between any pair of elements in a given graph. A typical graph is made up of a set of vertices (nodes) and edges (arcs), with the assumption that the graph's form is static [2]. A static graph can be an undirected graph, a directed graph, or a weighted graph (network). The three distinct dynamic graph structures are briefly listed below [24–26]:

- An undirected graph is a pair $G = (V, E)$, where V is a set of vertices, and E is a set of edges. Each edge is an unordered pair where v_i and $v_j \in V$;
- A directed graph is a pair $G = (V, A)$, where V is again a set of vertices, and A is a set of arcs. Each arc is an ordered pair (v_i, v_j), $i \neq j$;
- A weighted graph can be represented as $G = (V, E, W_V, W_E)$, where W_V represents the weight of vertices while W_E denotes the weight of edges. The weights can be in different styles: real numbers, complex numbers, integers, elements of some group, etc.

To characterize different graphs, a wide number of metrics were proposed in the literature, such as betweenness, out-closeness, and in-closeness [27]. The betweenness of a vertex v_j is the ratio of the distances between all pairs of other vertexes (v_i and v_k, $i \neq j \neq k$) that traverse v_j, to the total distance within the graph, as follows:

$$B_j = \sum_{i,k} \frac{d_{ijk}}{d_{ik}} \tag{2.1}$$

B_j represents the betweenness of vertex v_j; d_{ijk} denotes the distance between vertex v_i and vertex v_k that traverses vertex v_j; d_{ijk} denotes the distance between vertex v_i and vertex v_k that (does not) traverses vertex v_j.

In a directed graph, the closeness of a node v_j can be divided into two types: the out-closeness and the in-closeness. The out-closeness of a vertex v_j depends on the distance needed to reach from the vertex to other vertexes in the graph, as follows:

$$C_{\text{out},j} = \frac{1}{\sum_i d_{ji}} \tag{2.2}$$

$C_{out,j}$ represents the out-closeness of a vertex v_j; d_{ji} denotes the distance of the shortest path from v_j to v_i; The in-closeness of a vertex v_j is determined by the distance needed to reach the vertex from other vertexes in the graph, as follows:

$$C_{\text{in},j} = \frac{1}{\sum_i d_{ij}} \tag{2.3}$$

$C_{in,j}$ denotes the in-closeness of a vertex v_j; d_{ji} denotes the distance of the shortest path from v_i to v_j.

To model different systems, the distance can be defined by different parameters. For instance, d_{ij} may be represented by the ratio between the heat radiation threshold and the actual intensity of heat radiation from installation i to installation j. In that case, d_{ij} represents the escalation potential from installation i to installation j. The escalation potential d_{ij} increases with decreasing the propagation potential from installation i to installation j. The escalation potential of installation i increases with increasing the $C_{out,i}$ and the damage potential of installation i increases with increasing the $C_{in,i}$ [2, 28]. Consequently, graph metrics can be used to identify two types of critical vessels: (i) the critical vessels that have a high potential to initiate a fire-induced escalation and (ii) critical vessels with a high potential to be damaged by domino effects induced by other vessels nearby.

However, the graph metrics based on static graphs can not model the temporal characteristics of fire-induced escalations.

2.3.2 Dynamic Graphs

In real systems, graphs may change over time, such as computer programming languages and artificial intelligence. As a result, dynamic graphs were systematically proposed in the 1990s to model these practical dynamic applications. And some of the static graph algorithms were also improved to study the dynamic systems, such as the shortest path algorithms. A dynamic graph can also be an undirected graph, a directed graph, or a weighted graph. The structure of dynamic graphs is similar to that of static graphs, while the parameters in a dynamic graph may vary over time [24–26].

A dynamic graph G is updated when one or more of the following four graph parameters change: V (a set of vertexes), E (a set of edges), W_V (map vertexes to numbers), and W_E (map edges to numbers). According to the temporal characteristics, dynamic graphs can be divided into five types:

- Vertex dynamic graph: the set V changes over time, and the vertexes may be added or removed. When a vertex is removed or added, the related edges are also eliminated or increased;
- Edge dynamic graphs: the set E changes over time, and the edges may be added or removed. If an edge is removed or added, the related vertexes may also be deleted or added;
- Vertex weighted dynamic graphs: the weight W_V changes over time and the weights on the nodes update;
- Edge weighted dynamic graphs: the weight W_E changes over time, and the weights on the edge also update;
- Hybrid dynamic graphs: any combination of the above basic types can occur in real applications.

An operation that adds or eliminates vertexes or edges, or changes the weights of vertexes and edges, is referred to as an update on a graph. The graph can be considered unchanged between the two updates. As a result, a dynamic graph can be interpreted as a discrete series of static graphs, and each graph can be analyzed using static graph theory. Dynamic graph models can differ depending on the application, and similar algorithms can be built based on the dynamic graph's update laws [2, 24].

2.4 Domino Evolution Graph Model

Based on the theory of dynamic graphs and the characteristic of spatial–temporal evolution of domino effects, a new domino effect model is proposed in this section.

2.4.1 Dynamic Graph Definition

A Domino Evolution Graph (DEG) is a dynamic graph that represents the spatial–temporal evolution of domino effects triggered by intentional attacks or unintentional events. The DEG starts when there is a primary fire scenario and ends when the escalation is over. In this chapter, only a fire scenario is considered in the model, but it can be extended to model other escalation scenarios such as explosions and toxic dispersions. The DEG can be represented as a hybrid dynamic graph, as follows:

$$DEG = (N,\ E,\ W_N,\ W_E) \tag{2.4}$$

(1) N represents a set of nodes denoting hazardous installations in a chemical industrial area. In this chapter, only the hazardous installations with flammable materials are considered graph nodes since the dynamic graph model only addresses the escalation triggered by fire. The number of nodes (N) is considered unchangeable in the entire evolution process.

(2) E denotes a set of directed edges from installations inducing heat radiation to installations receiving the heat radiation. If an installation j is exposed to heat radiation caused by installation i, the two installations should be linked by a directed edge from node i to node j. Node i is often called tail, while node j is called head ($i \neq j$). An edge can be updated with the evolution of a domino effect.

(3) W_N represents a set of node weights. These weights characterize the vulnerability or harmfulness state of installations, as follows:

$$W_N = (S,\ Q,\ RTF,\ RTB,\ PP,\ PD) \tag{2.5}$$

(i) S represents a set of states that characterize the status of installations in an escalation process. According to the possible roles of hazardous installations in the evolution of fire-induced domino effects, three installation states are defined: "vulnerable", "harmful" and "failed". The description of these states is shown in Table 2.1. To better display the state of installations in a graph, installations in the "vulnerable" state are marked as yellow, those in the "harmful" state are marked as red, and those in the "failed" state are marked as gray in the dynamic graph. An escalation occurs when an installation in a "vulnerable state" transfers to a "harmful" state.

(ii) Q represents the total heat radiation received by installations, in kW/m^2. Only installations in the "vulnerable" state receive heat radiations from installations in a "harmful" state. The Q of installations in a "harmful" state or a "failed" state is considered to be zero.

Table 2.1 State descriptions (Chen et al. [2])

State	Description	Marked color
Vulnerable	The installation is not physically damaged, but it may receive heat radiation from other installations. The installation's temperature or internal pressure may increase in this state	Yellow
Harmful	The installation is on fire due to intentional events or due to escalation from other installations. Installations in this state have a harmful impact on other installations receiving their heat radiation	Red
Failed	The fire on the installation is extinguished due to the burning out of flammable substances or emergency response actions. All edges connected to the node will be removed if the installation's state transfers from "harmful" to "failed"	Gray

(iii) *RTF* denotes the residual time to failure (*RTF*) of installations in the "vulnerable" state, in min. *RTF* is equal to zero when an installation is damaged while it is assumed to be infinite when an installation is in the "failed" state.

(iv) *RTB* represents the residual time to burn out (*RTB*) of installations in the "harmful" state, in min. An installation in a "harmful" state transfers to a "failed" state when *RTB* is equal to zero.

(v) *PP* represents the probabilities of primary fire on installations. It denotes the vulnerability of installations directly against unintentional or intentional events. The *PP* may be decreased by improving the resistant capability of installations using safety and security measures. The likelihood of installations being damaged by domino effects is not considered in the weight.

(vi) W_E represents the weight of directed edges. It only consists of one weight: the heat radiations from tail installations to head installations, kW/m^2. The W_E can be represented by a square matrix of dimension $N \times N$ (an adjacent matrix), as follows:

$$EW = \begin{bmatrix} 0 & q_{12} & \cdots & q_{1n} \\ q_{21} & 0 & \cdots & q_{2n} \\ \cdots & \cdots & 0 & \cdots \\ q_{n1} & q_{n2} & \cdots & 0 \end{bmatrix} \tag{2.6}$$

where q_{ij} denotes the heat radiation intensity caused by installation *i* in the "harmful" state to installation *j* in the "vulnerable" state. q_{ij} is equal to zero when there is no directed edge from installation *i* to installation *j*. In the adjacent matrix, row *i* indicates the escalation potential of installation *i*, and column *j* represents the vulnerability of installation *j*.

2.4.2 Dynamic Graph Update

(1) Evolution update

A Domino Evolution Graph (DEG) consists of a chain of static graphs. The initial static graph (Graph 1) emerges when a primary scenario caused by unintentional or intentional events occurs. A new static graph will occur if an update operation is conducted. The graph index (g) is also updated according to the operation, as follows:

$$g = \begin{cases} 1 & \text{initial graph} \\ g+1 & \text{after a new update} \end{cases} \tag{2.7}$$

The time between two update operations is called "graph time" t (min). The evolution time at the beginning of graph g (T_g) can be obtained, as follows:

Fig. 2.2 State transition of installations (Chen et al. [2])

$$T_g = \begin{cases} 0 & g = 1 \\ T_{g-1} + t_{g-1} & g > 1 \end{cases} \tag{2.8}$$

(2) **State update**

Among the three installation states, update operations can be divided into two types: from "vulnerability" to "harmful" and from "harmful" to "failed", as shown in Fig. 2.2. In the first graph, the installations on fire are in the "harmful" state, and other installations are in the "vulnerable" state. If an installation is damaged and on fire due to escalation from external installations, its state will be changed from "vulnerable" to "harmful". Furthermore, if the fire on an installation is extinguished, the installation will be updated to a "failed" state. Finally, the update operation will terminate if no escalation occurs in which no installation is in the "vulnerable" state, and no installation is in the "harmful" state.

(3) **Directed edge update**

Directed edges link "harmful" installations with "vulnerable" installations. As a result, when the state of an installation is changed, the directed edges should be added or removed. When an installation's state changes to "harmful," all directed edges from other installations to the installation will be removed while all directed edges from the installation to other installations will be established. If an installation's state changes to "failed" the directed edges from (to) that installation to (from) other installations are erased.

(4) **Heat radiation update**

In a domino effect evolution process, installations with a "vulnerable" state can obtain heat radiations from more than one installation with "harmful" states, a phenomenon which is known as "synergistic effect". In contrast, an installation in the "harmful" state can emit heat to several installations in the "vulnerable" state, which is called "parallel effects". Figure 2.3a depicts a graph model of a parallel effect while Fig. 2.3b depicts a graph model of a synergistic effect.

Due to the synergistic effect, the total heat radiation suffered by an installation j in the "vulnerable state" (Q_j) should be the sum of heat radiations received from other installations in "harmful" states, as follows:

$$Q_j = \sum_{i=1}^{N} q_{ij} \tag{2.9}$$

The total amount of heat radiation obtained by each installation may change over time due to new occurrences of "harmful" installations or "failed" installations. The

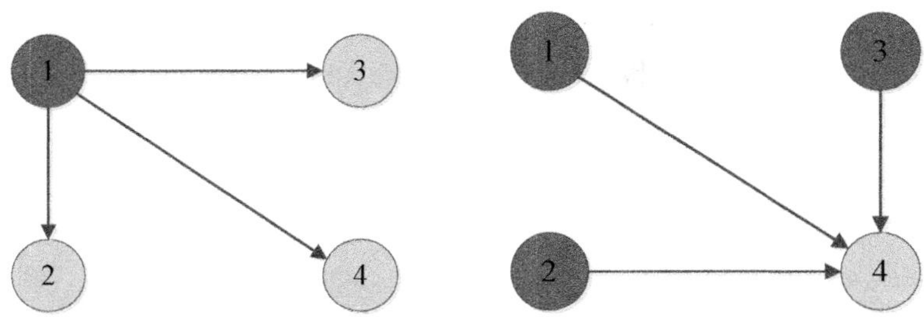

(a) Parallel (divergence) effects (b) Synergistic (convergence) effects

Fig. 2.3 Graph models of the spatial evolution of domino effects (Chen et al. [2])

heat radiation from installation i in the "harmful" state to installation j in the "vulnerable" state can be calculated by the application of consequence analysis software such as ALOHA [29].

(5) **Residual time to failure update**

Because of superimposed effects, the *RTF* of installations in "vulnerable" states may change over time in a domino effect evolution. Besides, passive protection measures such as fireproof coatings, have an impact on the *RTF*. When an installation j starts receiving heat radiation at evolution time T_g, the *RTF* can be obtained without considering the superimposed effect, as follows [8]:

$$RTF_{j,g} = \frac{\exp\left(c_1 \times V^{c_2} + c_3 \ln(Q_j) + c_4\right)}{60} \tag{2.10}$$

$RTF_{j,g}$ represents the residual time to failure of installation j at T_g (min); c_1–c_4 are parameters related to vessel factors, as shown in Table 2.2.

If $RTF_{j,g} < t_g$, the installation j will be physically damaged and escalation will occur at T_{g+1}. Otherwise, the *RTF* of installation j in the "vulnerable" state at the time T_{g+1} will be updated according to superimposed effects: the heat radiation in different stages received by the installation should be superimposed to calculate the *RTF* at the time of T_{g+1}, as follows [1]:

$$RTF_{j,g+1} = \left(\frac{Q_{j,g+1}}{Q_{j,g}}\right)^c \cdot \left(RTF_{j,g} - t_g\right) \tag{2.11}$$

(6) **Residual time to burn out update**

Table 2.2 The parameter value of *a, b, c,* and *d* (Landucci et al. [8])

Installation	c_1	c_2	c_3	c_4
Atmospheric tank	-2.67×10^{-5}	1	−1.13	9.9
Pressurized tank	8.845	0.032	−0.95	0

Given that an installation i is on fire at the time of T_g, *RTB* of the installation at that time can be obtained as the ratio of flammable material mass to burning rate, as follows:

$$RTB_{i,\,g} = \frac{m_i}{v_i} \tag{2.12}$$

m_i represents the mass of flammable substances in installation i (kg); v_i represents the burning rate of flammable substances in installation i (kg/min); $RTB_{i,\,g}$ denotes the *RTB* of installation i at the evolution time of T_g.

If $RTB_{j,\,g} < t_g$, the installation i will be extinguished; otherwise, it will continue to be on fire at T_{g+1}, and the *RTB* of installation i at T_{g+1} will be updated, as follows:

$$RTB_{i,\,g+1} = RTB_{i,\,g} - t_g \tag{2.13}$$

(7) **Escalation probability update**

In the chemical industry, emergency response is critical for preventing escalation or for mitigating domino effects [30]. As a result, emergency management should be considered when assessing the risk of plant installations. However, due to the complexities associated with human factors in executing emergency response activities, evaluating emergency response is very difficult. For the sake of simplicity, we assume that the domino effect progression can not be prevented until the emergency mitigation steps are effectively initiated [8]. In this chapter, a cumulative log-normal distribution (LND) function is used to model the time required to control domino effects (TTC) by emergency response actions, as follows:

$$\log TTC \sim N\left(u, \sigma^2\right) \tag{2.14}$$

u represents the mean of log *TTC* or expectation of the distribution; σ denotes the standard deviation of log *TTC* and σ^2 represents the variance. Maximum Likelihood Estimation (MLE) can be used to calculate these factors based on expert judgment, disaster drills, or models. Consequently, if an installation j is supposedly damaged at T_g with a certain probability during a domino effect evolution, the conditional probability of installation j being damaged (*PD*) given a primary scenario can be obtained, as follows:

$$P_{d,j} = \left(1 - \mathrm{LND}\left(T_g\right)\right) \tag{2.15}$$

2.5 Algorithm

The Domino Evolution Graph (DEG) model is developed in Sect. 2.4. This section introduces the Minimum Evolution Time (MET) algorithm [1] to solve the DEG model, as shown in Fig. 2.4. According to the algorithm, First, basic data needed for risk assessment of domino effects caused by fire need to be inputted, such as chemical plant layout information, potential heat radiations calculated by ALOHA software, and primary fire scenarios. Next, the parameters of the DEG model should be initialized given the primary fire scenario. Then, the *RTF* and *RTB* of each installation should be estimated according to the methods illustrated in Sect. 2.4. The initial DEG is updated when the evolution time is equal to the minimum value (t_m) of *RTF* and *RTB*, as follows:

$$t_m = \min(RTB_i, RTF_j)\; i, j = 1, 2, \ldots n \tag{2.16}$$

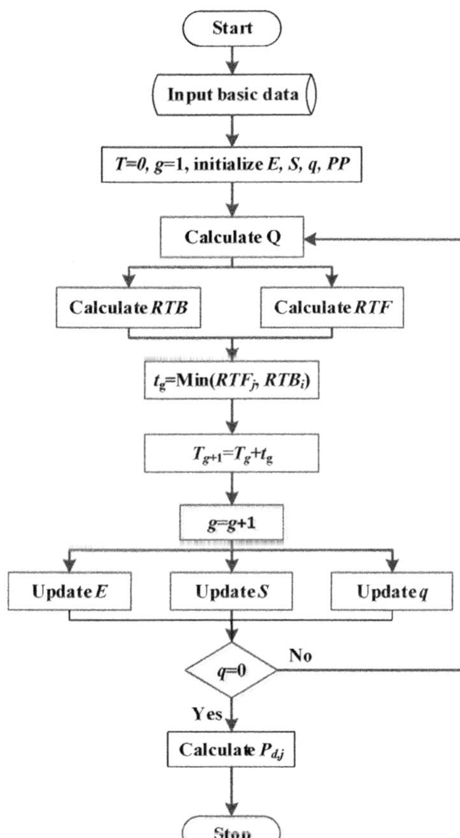

Fig. 2.4 Flow diagram of the MET algorithm for the DEG model (Chen et al. [2])

The parameters of E, S, q, and T are updated accordingly. The above steps will continue until q is equal to zero. Then, the graph update will terminate and the damage probability of each installation is calculated based on emergency response information. According to the algorithm, we can rapidly obtain the graph time, installation states at each time, the damage probability of installations, and the likelihood of domino effects.

2.6 Case Study

Figure 2.5 shows an illustrative chemical plant consisting of four storage tanks. The characteristics of these tanks are summarized in Table 2.3. The meteorological parameters of the chemical plant are assumed as follows: an ambient temperature of 20°C, the wind blowing from the West with a speed of 1.5 m/s, relative humidity of 50%, and the stability class D. The heat radiation intensity generated by a pool fire and the burning rate of each fire are calculated by the ALOHA software, as shown in Table 2.4. Assuming u is equal to 10 min and the corresponding variance is equal to 2 min [1].

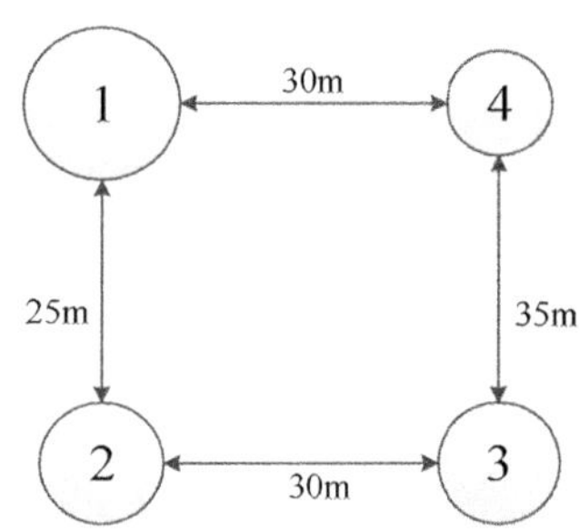

Fig. 2.5 Layout of an illustrative chemical plant (Chen et al. [2])

Table 2.3 Features of chemical storage tanks (Chen et al. [2])

Tank	Type	Dimension	Chemical substance	Volume (m^3)	Chemical content (t)
1	Atmospheric	30 × 10	Benzene	6000	4000
2	Atmospheric	20 × 10	Acetone	2500	2000
3	Atmospheric	20 × 10	Toluene	2500	1500
4	Atmospheric	10 × 6.5	Toluene	500	200

Table 2.4 The Heat Radiation from tank *i* to tank *j* and the *ttb* of tanks (Chen et al. [2])

Tank *i, j*	q_{ij} (kW/m^2)				*ttb* (min)
	1	2	3	4	
1	–	32.5	25.1	12.9	1666.7
2	17.7	–	13.2	4.1	1369.9
3	8.7	17.6	–	13.8	980.4
4	10.1	3.5	8.3	–	233.9

2.6.1 The Application of Graph Metric Approach

Using the static graph methodology, setting a threshold value of 15 kW/m^2, a static graph of the chemical plant is obtained, as shown in Fig. 2.6. The calculation results of the graph metrics (e.g., out-closeness and in-closeness) of the four tanks are listed in Table 2.5.

According to the out-closeness results illustrated in Table 2.5, Tank 1 has the highest potential to initiate a fire-induced domino effect if it is on fire, followed by Tank 2 and then Tank 3. The fire at Tank 4 can not trigger a domino effect. Based on the in-closeness value, Tank 2 is most vulnerable to domino effects, followed by Tank 3, and also Tank 1. Tank 4 is more likely not to fail when a domino effect occurs at one of the other tanks in the chemical plant.

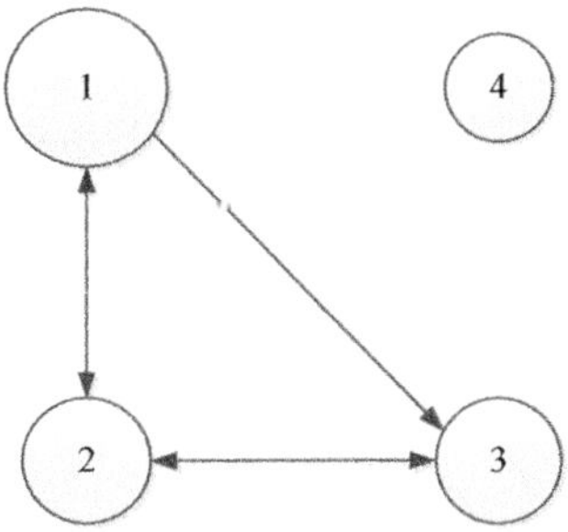

Fig. 2.6 Static graph model of the chemical plant (Chen et al. [2])

Table 2.5 The results of closeness for the static graph in Fig. 2.7 (Chen et al. [2])

Tank	Out-closeness	In-closeness
1	0.42	0.17
2	0.19	0.34
3	0.17	0.22
4	0	0

Table 2.6 The damage time of tanks (min) (Chen et al. [2])

Tank	S1	S2	S3	S4	S5
1	0	11.0	19.2	–	0
2	6.1	0	12.2	–	0
3	7.4	16.1	0	–	5.1
4	13.5	20.3	22.2	0	9.3

2.6.2 *The Application of Dynamic Graph Approach*

According to the dynamic graph approach, five primary scenarios are assumed, as follows: (S1) Pool fire at Tank 1; (S2) Pool fire at Tank 2; (S3) Pool fire at Tank 3; (S4) Pool fire at Tank 4; (S5) Pool fires at Tank 1 and Tank 2. Using the dynamic graph approach, the failure time of each tank caused by different primary scenarios can be estimated, as shown in Table 2.6.

Taking primary scenario 2 as an example, Tank 1 is on fire at $T = 0$ min (Fig. 2.7a). The heat radiation emitted from Tank 2 can cause credible damage to tanks 1, 3, and

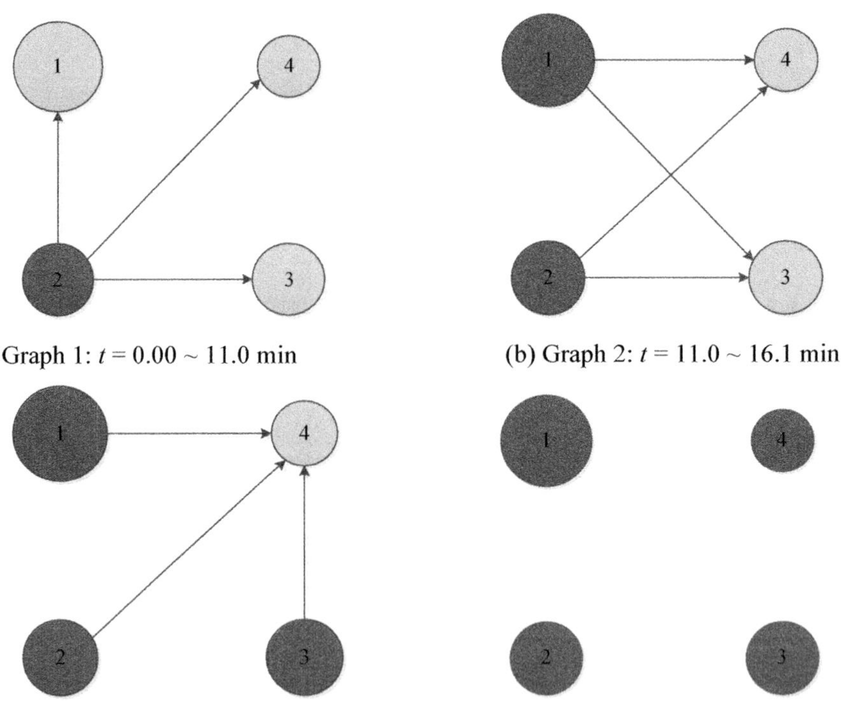

(a) Graph 1: $t = 0.00 \sim 11.0$ min (b) Graph 2: $t = 11.0 \sim 16.1$ min (c) Graph 3: $t = 16.1 \sim 20.3$ min (d) Graph 4: $t = 20.3 \sim 254.2$ min

Fig. 2.7 The DEG caused by a fire at tank 2 (Chen et al. [2])

Table 2.7 The conditional probability of tanks being damaged (Chen et al. [2])

Tank	A1	A2	A3	A4	A5	PBD
1	1.00	0.23	1.34×10^{-6}	0	1.00	0.45
2	1.00	1.00	0.07	0	1.00	0.61
3	0.98	2.95×10^{-4}	1.00	0	1.00	0.60
4	0.01	1.70×10^{-7}	4.91×10^{-9}	1.00	0.68	0.34
PCD	0.75	0.31	0.27	0.25	–	–

4, leading to a fire at Tank 1 at $T = 11.0$ min. In that case, the state of Tank 1 transfers from "vulnerable" to "harmful", leading to a synergistic effect on Tank 3 and Tank 4, as shown in Fig. 2.7b. Then, Tank 3 is on fire at $T = 16.1$ min (Fig. 2.7c) due to the synergistic effect and a superimposed effect caused by Graph 1 and Graph 2. Finally, Tank 4 catches fire at $T = 20.3$ min.

Considering the uncertainty of emergency response according to Eq. (2.14), we can obtain the damage probability of each Tank of different primary scenarios, as shown in Table 2.7. It indicates that the domino effect caused by the attack on Tank 1 may be inevitable due to the fastest evolution. Tank 2 with the highest average conditional probability of being damaged (PBD) is more susceptible to domino effects caused by other tanks. While Tank 1 has the highest potential to initiate domino effects since it has the largest value of the average conditional probability to cause damage (PCD).

2.7 Discussion

In this study, we introduced the static graph method and the dynamic graph method for modeling fire-induced domino effects. According to the results of a case study, the ranking of tanks based on their in-closeness values is the same as their ranking based on their average conditional probability of being damaged (PBD). Besides, the ranking of tanks based on their out-closeness is also identical to their ranking based on their average conditional probability to cause damage (PCD). Therefore, It may be demonstrated that the DEG model proposed in this chapter is valid. Moreover, The dynamic graph approach can obtain the dynamic evolution of domino effects compared to the static graph which seems to provide merely a snapshot of the whole process at once.

This study develops the Minimum Evolution Time (MET) algorithm to solve the DEG model, avoiding the time-consuming approach of repeatedly generating random numbers. The maximum number of iterations of the MET algorithm is $2n$ while the minimum number of iterations of Monte Carlo-based simulation tools is 10^4–10^6 [17, 20]. So the calculation efficiency is improved greatly. As a result, this approach can be used for assessing fire-induced domino effects in chemical clusters with a large number of hazardous installations.

In the DEG model, an adjacent matrix is used to represent potential heat radiations between each pair of tanks. In that case, the primary scenario with multiple fires that is more likely to be induced by multiple attacks can be addressed. As shown in Table 2.6, both Tank 1 and Tank 2 are on fire in primary scenario 5, leading to a faster evolution than the primary scenarios with single fires (S1, S2, S3, S4, and S5) and making the prevention of a domino effect more difficult. Besides, parallel effects, synergistic effects, and superimposed effects can also be considered in this model.

In this chapter, we only consider fire and neglect vapor cloud explosion (VCE) in domino effects. The next chapter will model domino effects triggered by VCEs.

2.8 Conclusions

This chapter introduces a Domino Evolution Graph (DEG) model and a Minimum Evolution Time (MET) algorithm to model the spatial–temporal evolution of domino effects triggered by fires. The evolution process is divided into stages according to chronological order. A series of static graphs representing a domino evolution process is sequentially connected by superimposed effects, constituting a DEG. To rapidly solve the DEG model, the MET algorithm based on the principle of minimum evolution time is developed. By applying the MET algorithm, we can obtain the evolution graphs, evolution time, evolution probability, escalation probability, and the damage probability of each installation exposed to fire-induced domino effects. Compared with graph metrics, the DEG model can grasp the dynamic evolution characteristics of domino effects and overcome the limitation of the "probit model" in higher-level escalations. The developed model can be applied to support the decision-making on safety and security barriers since it can be used in real chemical plants and chemical clusters with a large number of hazardous installations. Besides, this approach can also be extended to domino effect assessment related to multi-hazard accident scenarios (fire, explosion, and toxic release).

References

1. Chen C, Reniers G, Zhang L (2018) An innovative methodology for quickly modeling the spatial-temporal evolution of domino accidents triggered by fire. J Loss Prev Process Ind 54:312–324
2. Chen C, Reniers G, Khakzad N (2019) Integrating safety and security resources to protect chemical industrial parks from man-made domino effects: a dynamic graph approach. Reliab Eng Syst Saf 191
3. Chen C (2021) A dynamic and integrated approach for modeling and managing domino-effects. Delft University of Technology
4. Chen C, Khakzad N, Reniers G (2020) Dynamic vulnerability assessment of process plants with respect to vapor cloud explosions. Reliab Eng Syst Saf 200
5. Darbra RM, Palacios A, Casal J (2010) Domino effect in chemical accidents: main features and accident sequences. J Hazard Mater 183(1–3):565–573

6. Necci A, Cozzani V, Spadoni G, Khan F (2015) Assessment of domino effect: state of the art and research needs. Reliab Eng Syst Saf 143:3–18
7. Alileche N, Cozzani V, Reniers G, Estel L (2015) Thresholds for domino effects and safety distances in the process industry: a review of approaches and regulations. Reliab Eng Syst Saf 143:74–84
8. Landucci G, Gubinelli G, Antonioni G, Cozzani V (2009) The assessment of the damage probability of storage tanks in domino events triggered by fire. Accid Anal Prev 41(6):1206–1215
9. Reniers GLL, Dullaert W (2007) DomPrevPlanning©: User-friendly software for planning domino effects prevention. Saf Sci 45(10):1060–1081
10. Reniers G, Dullaert W, Audenaert A, Ale BJ, Soudan K (2008) Managing domino effect-related security of industrial areas. J Loss Prev Process Ind 21(3):336–343
11. Zhang X, Chen G (2011) Modeling and algorithm of domino effect in chemical industrial parks using discrete isolated island method. Saf Sci 49(3):463–467
12. Khakzad N, Reniers G (2015) Using graph theory to analyze the vulnerability of process plants in the context of cascading effects. Reliab Eng Syst Saf 143:63–73
13. Khakzad N, Reniers G, Abbassi R, Khan F (2016) Vulnerability analysis of process plants subject to domino effects. Reliab Eng Syst Saf 154:127–136
14. Khakzad N, Reniers G, van Gelder P (2017) A multi-criteria decision making approach to security assessment of hazardous facilities. J Loss Prev Process Ind 48:234–243
15. Khakzad N, Khan F, Amyotte P, Cozzani V (2013) Domino effect analysis using Bayesian networks. Risk Anal 33(2):292–306
16. Khakzad N (2015) Application of dynamic Bayesian network to risk analysis of domino effects in chemical infrastructures. Reliab Eng Syst Saf 138:263–272
17. Zhang L, Landucci G, Reniers G, Khakzad N, Zhou J (2017) DAMS: a model to assess domino effects by using agent-based modeling and simulation. Risk Anal
18. Zhou J, Reniers G (2016) Petri-net based simulation analysis for emergency response to multiple simultaneous large-scale fires. J Loss Prev Process Ind 40:554–562
19. Zhou J, Reniers G, Khakzad N (2016) Application of event sequence diagram to evaluate emergency response actions during fire-induced domino effects. Reliab Eng Syst Saf 150:202–209
20. Abdolhamidzadeh B, Abbasi T, Rashtchian D, Abbasi SA (2010) A new method for assessing domino effect in chemical process industry. J Hazard Mater 182(1–3):116–126
21. Gómez-Mares M, Zárate L, Casal J (2008) Jet fires and the domino effect. Fire Saf J 43(8):583–588
22. Reniers G, Cozzani V (2013) Domino effects in the process industries, modeling, prevention and managing. Elsevier, Amsterdam
23. Khakzad N, Khan F, Amyotte P, Cozzani V (2014) Risk management of domino effects considering dynamic consequence analysis. Risk Anal 34(6):1128–1138
24. Harary F, Gupta G (1997) Dynamic graph models. Math Comput Model 25(7):79–87
25. Bondy JA, Murty USR (1976) Graph theory with applications, vol 290. Citeseer
26. Casteigts A, Flocchini P, Quattrociocchi W, Santoro N (2012) Time-varying graphs and dynamic networks. Int J Parallel Emergent Distrib Syst 27(5):387–408
27. Hernández JM, Van Mieghem PJDUoTM, The Netherlands (2011) Classification of graph metrics 1–20
28. Khakzad N (2018) A graph theoretic approach to optimal firefighting in oil terminals. Energies 11(11)

29. ALOHA user manual (2016) US Environmental Protection Agency, National Oceanic and Atmospheric Administration, ALOHA, Version 5.4.7. https://www.epa.gov/cameo/aloha-software. Accessed 5 May 5 2019
30. Zhou J, Reniers G (2016) Petri-net based modeling and queuing analysis for resource-oriented cooperation of emergency response actions. Process Saf Environ Prot 102:567–576

Chapter 3
Dynamic Assessment of VCE-Induced Domino Effects

3.1 Introduction

In the process and chemical industry,[1] a release of hazardous materials may lead to a fire if the released materials are immediately ignited, while it is more likely to induce a VCE when the ignition is delayed. Fire, explosion, and toxic dispersion induced by a loss of containment (LOC) are defined as major accident scenarios [4–6]. Among these major accident scenarios, the most likely scenario is fire, while explosions may lead to damage in a wider area and have serious consequences, resulting in deaths and significant collateral loss [7]. Due to the vaporization of liquid materials, large releases usually induce vapor cloud explosions (VCE) rather than fires [8]. Both fire and explosion are common escalation scenarios that can trigger domino effects. Abdolhamidzadeh et al. [9] investigated 224 domino accidents in the process industry and demonstrated that explosion is the most frequent cause of domino effects (57%). Besides, VCE accounted for 84% of the domino effects induced by explosions. It should be noted that, in recent years, several domino effects by VCEs led to catastrophic consequences, such as the Puerto Rico explosion (2009, USA), the Sitapura explosion (2009, India), and the Amuay explosion (Venezuela, 2012).

Although many VCE-induced domino effects occurred in the process and chemical industry, the mechanism of the blast with large shock waves is not well understood [10]. For example, no consistent interpretation is obtained for the VCE-induced domino effect in the Buncefield depot in the UK, which brought about an unexpected overpressure with a maximum value of more than 2000 kPa [11]. In a VCE event, a fast expansion of flames can generate shock waves or so-called overpressures, which may damage hazardous installations nearby, leading to a domino effect accident [12]. Many factors influence the expansion of flames and the intensity of overpressures, such as the type and quantity of the released flammable materials, the delayed time

[1] This chapter is mainly based on three publications: Chen et al. [1, 2] and Chen [3].

C. Chen et al., *Integrating Safety and Security Management to Protect Chemical Industrial Areas from Domino Effects*, Springer Series in Reliability Engineering,
https://doi.org/10.1007/978-3-030-88911-1_3

to ignition (DTI), the space figuration of the release position and the position and the number of ignition sources [13].

Many attempts have been conducted in the academic domain to model flammable gas dispersion or estimate the intensity of overpressure generated by VCEs. The TNT equivalent method [14] is the most widely used in risk analysis [15, 16] since it is a user-friendly solution for the estimation of a far-field blast effect. But it may underestimate the intensity of overpressure due to the neglection of the space configuration, ignition sources, and the dispersion of the vapor cloud [17, 18]. Compared with the equivalent method, the Multi-Energy method considers more the gas explosion mechanisms in which a VCE event is modeled by multiple sub-explosions inside specific obstructed areas. The Multi-Energy method is recommended as an alternative for the TNT equivalent method in quantitative risk analysis [12]. Past research indicates that this method is more suitable for estimating large shock waves produced by long-delayed VCEs in fuel storage plants [11, 19, 20]. Besides, the Baker-Strehlow method [21] and CFD simulation [22, 23] are common-used methods for calculating overpressure intensity.

Consequently, VCE-induced domino effects may be more challenging to assess due to the uncertainty of ignition positions, the uncertainty of delayed ignition time (DIT), and the complexity of overpressure intensity calculations. The VCE induced by a LOC of hazardous materials in a chemical plant is time-dependent due to vapor cloud dispersion and delayed ignition. However, previous risk analysis methods [24, 25] for VCE or VCE-induced domino effects always consider that the explosion occurs immediately at the release point, which is inconsistent with the observations from past VCE accidents in the process and chemical industry. The explosion position is related to the vapor cloud dispersion and ignition source distribution.

Consequently, a dynamic risk assessment methodology based on a discrete dynamic event tree (DDET) is introduced in this chapter. The developed method integrates individuals, hazardous installations and ignition sources into a stochastic simulation engine to model the time dependencies and ignition uncertainties in the evolution of VCEs. In the methodology, the shock waves induced by VCEs are calculated by the Multi-Energy method, while the damage probability of installations is estimated using probit models. VCE, its characteristics, and previous studies are represented in Sect. 3.2. Section 3.3 introduces the dynamic methodology. Finally, conclusions are drawn in Sect. 3.4.

3.2 VCE-Induced Escalations

This section analyzes past VCE accidents in the process and chemical industry and summarized the mechanisms of VCEs, calculation methods of overpressure intensity, and the methods to assess the escalation induced by VCEs.

3.2.1 Past VCE Accidents

In light of the frequency and severe consequences, VCEs have gained considerable attention in recent years [19]. Table 3.1 shows several well-known VCE-induced domino accidents in the most recent two decades in the process and chemical industry.

The Skikda accident in 2004

On the evening of January 19, 2004, a blast occurred at a liquefied natural gas (LNG) plant near the eastern Algerian port of Skikda. A gas pipeline leak (Boiler explosion) was deemed the cause of the explosion. The Skikda LNG accident led to 27 deaths of employees, 56 injures, and a $900 million property loss. The explosion caused a subsequent domino effect, resulting in serious damage to three of the six LNG trains at the plant and a nearby berth for loading LNG tankers. The Skikda LNG accident was the worst petrochemical plant accident in Algeria. The unreasonable layout in which hazardous installations were located nearby, made the storage plant more vulnerable to domino effects [26].

The Buncefield accident in 2005

On the morning of December 11, 2005, a series of explosions occurred in the Buncefield storage and transfer depot, Hemel Hempstead, the U.K., resulting in the damage of large parts of the depot and neighboring properties and causing 43 injuries [27]. The accident was caused by an overfilling when a storage tank received an excessive amount of unleaded gasoline at about 5:20 a.m. A massive explosion and following explosions occurred at 06:01 a.m. when the vapor cloud was ignited, creating much higher overpressures than those estimated by methods usually applied [11]. These explosions caused subsequent fires, which engulfed 20 out of 39 large storage tanks. The fire lasted for five days, destroying most facilities and causing 2000 people to evacuate [2].

Table 3.1 A summary of past VCE events in the past two decades (Chen et al. [1])

Event	Plant type	Ignition source	DIT (s)
Skikda, Algeria 2004	LNG facility	Boiler explosion	<300
Buncefield, UK 2005	Gasoline storage	Pumphouse	1380
San Juan, Puerto Rico 2009	Gasoline storage	Wastewater treatment area	1560
Jaipur, India 2009	Gasoline storage	Control room	4500
Amuay, Venezuela 2012	Refinery LPG storage	Vehicle	4080
Zhangjiakou, China 2018	Chemical plant	Furnace	418

The Puerto Rico accident in 2009

On the night of 23 October 2009, a domino effect accident was seen, at a storage plant of Caribbean Petroleum Corporation, Puerto Rico, in the U.S. This accident is also caused by an overfilling when unleaded gasoline was unloaded from a cargo ship to various tanks in the plant. The unloading operation began at 8:47 p.m., when the gasoline was first pumped to tank 405. Operators fully opened the value at Tank 409 after tanks 405, 504, and 411 were filled. Then the tank began to overflow between 11 p.m. and 12 a.m. The overfilling created a large vapor cloud that caught fire, resulting in a VCE and subsequent fires. The accident destroyed 17 out of 48 tanks and other equipment onsite and damaged neighborhood communities and businesses. The fires lasted 60 h and induced an emergency declaration for assistance by President Obama [2, 28].

The Jaipur accident in 2009

On the night of October 29, 2009, only one week after the Puerto Rico accident, a similar VCE-induced domino effect accident occurred in the Indian Oil Corporation refinery, 16 miles south of Jaipur, India. A delivery pipeline leaked when four operators were transferring gasoline to a storage tank, resulting in a large vapor cloud that covered a major area of the plant. The release lasted about 75 min before the vapor cloud was ignited, resulting in a powerful explosion and subsequent fireballs and fires. The emergency measures in the plant never took place since the operators involved had been overcome by the fire. Consequently, the fires could not be extinguished and burned for 11 days. It was estimated that over 2000 tons of gasoline were released, and 4% of the released gasoline was vaporized and formed a vapor cloud. Finally, the accident destroyed the entire site and caused 11 fatalities and over 200 injuries [2, 29].

The Amuay refinery accident in 2012

On the morning of August 25, 2012, an explosion occurred at the Punto Fijo (Amuay) refinery in Paraguana, the northwest of Venezuela. It was the world's second-largest refinery in which tons of crude oil, liquefied petroleum gas (LPG), and liquefied natural gas (LNG) are stored and handled. At 12:35 a.m., firefighters received the first alert on the release of propane in the storage area and responded immediately by cordoning off the area and closing the Bolivar Avenue Judibana and Av. at 1:07 a.m., a large vapor cloud formed and ignited, resulting in a large-scale shock wave, resulting in the destruction of homes and surrounding structures and Detachment No. 44 of the National Guard. The accident led to over 50 deaths and more than 100 severe injuries, and several light injuries.

The Zhangjiakou accident in 2018

On the early morning of November 28, 2018, a vapor cloud explosion occurred outside a chemical plant at Zhangjiakou, China, leading to 24 fatalities and 22 injuries [30]. The accident was caused by a Vinyl chloride release from the vinyl chloride gas tank of Hebei Shenghua Chemical Co., Ltd. of China National Chemical Corporation.

The released vinyl chloride spread to the road outside the plant and deflagrated in the event of an open flame, resulting in the death of 24 people, including the drivers of the vehicles parked on both sides of the road waiting for unloading (including 1 death in the later treatment) and 22 injuries. This accident evidenced that VCE risk could be underestimated if we ignore the uncertainty of explosion positions. VCEs can occur both inside and outside the process plant where the release occurs [30].

3.2.2 Explosion Mechanism

During the leakage of flammable materials, a flammable vapor cloud (FVC) can be created by mixing released flammable gases or evaporated flammable liquids with air. The potential outcomes of vapor cloud ignition are flash fire (FF) and VCE. Suppose the concentration of flammable gases is within the flammability limits (between the lower flammability limit and the upper flammability limit) and an ignition source is present to provide the necessary energy. In that case, ignition can occur (usually of the order of 10 J). FFs can be created by the abrupt combustion of an FVC, where the flame is not intensified due to a lack of obstacles or the effect of turbulent dispersion. Alternatively, the blaze speed could boost to sufficiently high velocities, resulting in severe blast overpressure [13].

A VCE's expansion mechanism can be explained using the flame speed proportional to the formed blast waves. Following an explosion, the flame begins to spread away from the point of ignition at a rate of 5–30 m/s, causing relatively slight overpressure. Owing to the chaotic existence of the flame and massive turbulent eddies, a wrinkled-flame front emerges, resulting in a rise of the flame surface and the acceleration of flame speed (30–500 m/s.) and the formation of an overpressure of up to 2–3 mbar. The presence of barriers in the flow causes the flame speed to rise (500–1,000 m/s), resulting in an overpressure of up to 1 bar. This physical mechanism of the flame speed is known as deflagration. When the reactive mixture's temperature is higher than the self-ignition temperature, and the reactive mixture in the front zone of turbulent combustion is compressed and heated due to mixing with combustion materials, a shock wave can be formed. This physical reaction is known as detonation, and it can result in flame speeds of up to 2,200 m/s and overpressures of up to 20 bar. Johnson and Tam [31] demonstrated that the large overpressure induced by VCEs in industrial plants can be explained by the theory of deflagration to detonation transition (DDT). Atkinson et al. [32] deemed that episodic deflagrations may be responsible for very large VCEs due to natural flame instability. Consequently, the VCE mechanisms need to be further studied in the future since there is no consistent statement to explain the VCEs in open areas.

3.2.3 *Impact Assessment of Vapor Cloud Explosions*

Empirical analytic models and computational models are two types of models that can approximate or forecast the results of vapor cloud explosions. In risk assessment, empirical analytic models are always used, such as the congestion measurement method, TNT Equivalent method, Multi-Energy method, and Baker-Strehlow method [15]. Computational models are mainly based on CFD codes, such as the commercial software of Flacs and Fluent [23, 33].

(1) **TNT Equivalent method**

The most widely used approach is the TNT equivalent method. The available combustion energy of a vapor cloud is converted into an equivalent charge weight of TNT via an efficiency factor [14]. This method provides a simple method for estimating the far-field blast effect, neglecting the space configuration where the explosion occurs [17, 18].

(2) **Multi-Energy method**

The Multi-Energy approach is based on a gas explosion mechanism that considers the VCE as a sequence of sub-explosions within special obstructed areas [12]. In this approach, the layout of the space where the cloud is spreading is represented as a strength coefficient. The value of the coefficient is proportional to the blast overpressure, which increases as the obstacle density in the region increases. In general, the TNT equivalent method can be easily used to measure overpressure, while the Multi-Energy method is more precise and similar to real conditions [13].

(3) **Baker-Strehlow method**

The Baker-Strehlow method, which is built on the Multi-Energy method, considers the flame propagation speed [21]. The speed of flame propagation is determined by how the flame front propagates, the reactivity of the fuel, and the density of the obstacles [15].

(4) **Numerical models**

In recent years, numerical approaches focused on CFD codes have attracted a lot of interest [22, 23, 34]. These methods can simulate the effects of terrain shape and the presence of obstacles on vapor cloud dispersion [35]. CFD simulation of the entire manufacturing facility can provide more precise outcomes. Still, it is very complicated, time-consuming, and costly, making it unsuitable for risk analysis of massive and complex production plants [13]. The accuracy of CFD codes, particularly in the case of VCE simulation in congested environments, is determined by the combustion models, turbulence closure models, constants for turbulence computation, and the definition of the dynamic relationship between fame front and turbulent flow area [34].

3.2.4 *Frequency Assessment of Vapor Cloud Explosions*

In the process and chemical industries, event trees have been commonly used to evaluate scenarios and the probability of incidents triggered by a LOC event [25, 36–39]. A general event tree analysis for the release of hazardous liquid substances from an atmospheric storage tank is shown in Fig. 3.1.

If the released material is ignited directly after a LOC incident, a pool fire scenario will occur. Otherwise, the material emitted will vaporize and create a vapor cloud. A flammable vapor cloud (FVC) can induce a VCE or FF during the dispersion process in the event of a delayed ignition. If there is no immediate ignition and the ignition is postponed, the release event may create a large hazardous vapor cloud that may be dangerous to nearby people or the environment.

The assessment of fire injuries caused by instant combustion is fair since there is no delay, and the fire should be considered to occur at the release position. However, it ignores the uncertainty of delayed ignition time (DIT), the uncertainty of ignition, and the uncertainty of delayed ignition location, all of which are essential for evaluating VCEs. First, the composition of the vapor cloud grows over time, which has a significant effect on the probability of combustion and the duration of the blast. The ignition location, on the other hand, affects the severity of the blast as well as the damage effects on other installations. As shown in Table 3.1, DIT values vary from 20 to 4500 s, and ignition source areas may be located within (e.g., pump house, wastewater treatment areas) or outside (e.g., vehicles) chemical plants. Consequently, failing to account for the uncertainty caused by vapor cloud dispersion can result in substantial errors. Furthermore, most of these incidents happened in no-wind or low-wind conditions, suggesting that stable and large vapor clouds are more likely to form in these conditions [32, 40].

3.3 Dynamic Vulnerability Assessment Methodology

Despite extensive research into the susceptibility of facilities to vapor cloud explosions triggered by LOC, the spatial–temporal evolution of such incidents has been

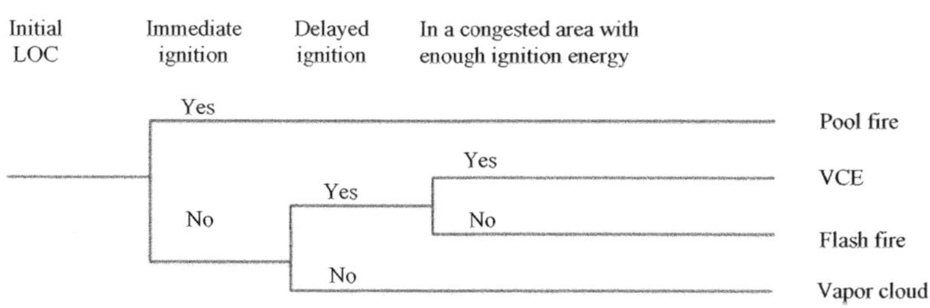

Fig. 3.1 An event tree for analysis of LOC-induced accident scenarios (Chen et al. [1])

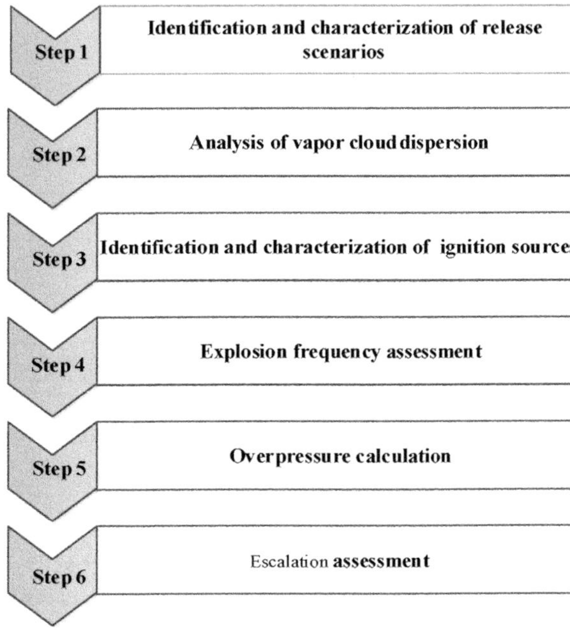

Fig. 3.2 Flow chart of DVEA methodology (Chen et al. [1])

underestimated. The susceptibility of installations, on the other hand, is determined by the degree of overpressure induced by the VCE, which is important to the spatial–temporal evolution of vapor clouds before ignition. Thus, this segment aims to develop a dynamic VCE evolution assessment (DVEA) methodology that integrates the vapor cloud dispersion mechanism and ignition uncertainty into a stochastic simulation engine to assess the vapor cloud explosion risk in chemical industrial areas. Figure 3.2 shows the flow chart of the DVEA methodology. The following paragraphs go through the methodology's different steps in greater detail.

3.3.1 Step 1: Identification and Characterization of LOC Scenarios

Significant volumes of volatile (flammable/explosive/toxic) compounds are treated, shipped, and processed in process industrial areas using various installations such as process pipes, pipelines, valves, flanges, heat exchangers, pumps, storage tanks, and so on. An installation's intrinsic hazard is determined by the volume of material present, the dangerous properties of the substance, and the relevant operating conditions (e.g., temperature, pressure) [41]. Since this research explicitly deals with fire and explosion incidents, other adverse consequences of dangerous chemicals are excluded. As a result, the first phase of the established approach recognizes only the hazardous installations that can become a source of flammable or explosive substances. Following the identification of potentially dangerous facilities, the failure of containment (LOC) incidents should be characterized to assess the LOC situations and the occurrence of each scenario. In this stage, all LOCs triggered by accidental events and deliberate events should be defined. The former should include generic LOCs, external-impact LOCs, loading and unloading LOCs, and other types of LOCs [12].

Generic LOCs cover all failure factors that are not explicitly addressed, such as corrosion, fabrication defects, welding faults, and tank vent blockage. External-impact LOCs are designed primarily for transportation units. LOCs that exist during loading and unloading operations are referred to as loading and unloading LOCs, such as overfilling [2]. In intentional LOCs (for example, intentionally opening valves), a security risk analysis based on available intelligence should be performed, including hazard analysis, attraction analysis, vulnerability analysis, effect analysis, etc. [42–45]. The metrics used for vapor cloud analysis, such as the initial pressure and temperature of hazardous substances in buildings, the mass of hazardous substances, and the potential leak sizes, can be defined based on the analysis of LOC scenarios.

3.3.2 *Step 2: Analysis of Vapor Cloud Dispersion*

The results of hazardous installation identification and LOC scenario characterization, such as release location, maximal release period, and mass flow rate, are needed for vapor cloud formation and dispersion analysis. As a result, this step helps to model the vapor cloud dispersion phase over time, resulting in the vapor cloud volume and location over time. Considering the initial release time is zero, the total released mass (M_t) at time t for a time-varying release scenario can be expressed as the integral of the mass flow rate (m_t) for time, as follows:

$$M_t = \int_0^t m_t \mathrm{d}t \tag{3.1}$$

m_t represents the mass flow rate which is a function of time t. To simplify the calculation, the period t can be divided into n discrete segments. In that case, the M_t can approximately be the sum of the masses in each segment, as follows:

$$M_t = M_{t_1} + M_{t_2} + M_{t_3} + \cdots + M_{t_n} \tag{3.2}$$

If the mass flow rate does not depend on time and it can be considered a constant (m), then Eq. (4.2) can also be further simplified, as follows:

$$M_t = m \times t \tag{3.3}$$

Considering the leak is a hole, the mass flow rate m of the leakage can be obtained as follows [11]:

$$m = \begin{cases} C_d A_h P_o K \sqrt{\frac{W_g}{\gamma RT}} \text{ leakage of gases} \\ C_d A_h \sqrt{2(P - P_a)} \text{ leakage of liquids} \end{cases} \tag{3.4}$$

C_{d} denotes the discharge coefficient; A_h (m^2) represents the cross-sectional area of the leakage hole; P_{o} (Pa) represents the initial gas pressure in the vessel (for each time step); W_{g} (kg/mol) represents the molecular weight of the gas; γ represents the Poisson ratio; R represents the universal gas constant (8.314 $\mathrm{Jmol^{-1}\ K^{-1}}$); T (K) represents the temperature of the gas; Pa (Pa) denotes the ambient pressure; ρ ($\mathrm{kg/m^3}$) represents the density of the liquid.

The mass of the flammable substance in the vapor cloud $M_{f,\mathrm{t}}$ is expressed as the released mass multiplied by the ratio of the evaporation rate to flow rate α. α is equal to 1 if the released material is a gas. The volume of the flammable gas ($V_{f,t}$) can be obtained as follows:

$$V_{f,t} = \frac{\alpha \times M_{f,t}}{\rho_f} \tag{3.5}$$

ρ_f represents the density of the flammable gas. A vapor cloud is a mixture of air and flammable vapor. As a result, the total volume of the vapor cloud (V_t) can be obtained as follows:

$$V_t = V_{f,t} + V_{a,t} \tag{3.6}$$

$V_{a,t}$ represents the volume of air mixed in the vapor cloud, which can be determined by considering that the flammable gas is thoroughly mixed with oxygen [13]. In the Multi-Energy method, the vapor cloud is modeled as a hemisphere shape [14]. Figure 3.3 shows a sketch of the vapor cloud hemisphere model in a chemical plant. This model considers possible hazardous installations, ignition sources, and obstacles covered by the vapor cloud. In that case, the boundary of the vapor cloud can be characterized by the radius of the hemisphere (R_t), as shown in Eq. (3.7).

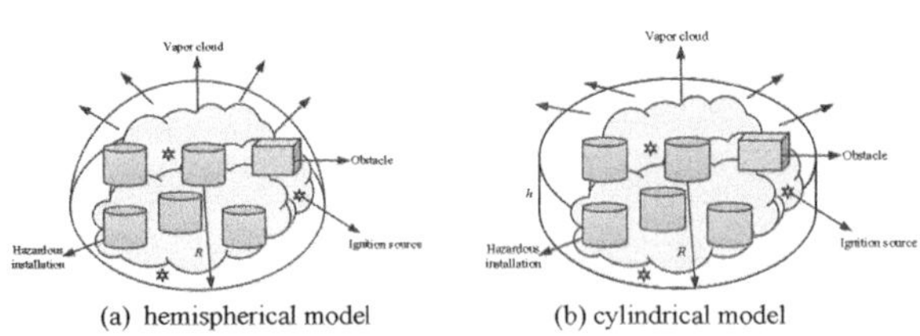

Fig. 3.3 A sketch of vapor cloud dispersion in a chemical industrial plant (Chen et al. [1])

$$R_t = \left(\frac{3V_t}{2\pi}\right)^{\frac{1}{3}} \tag{3.7}$$

If the released material is a dense gas, a cylindrical shape can be suitable to model the dispersion process, as shown in Fig. 3.3b [12, 46, 47]. A consistent height h and a radius of R are used to characterize the dispersion, as follows:

$$R_t = \left(\frac{V_t}{h\pi}\right)^{\frac{1}{2}} \tag{3.8}$$

In actual cases, the height of the vapor cloud varies with the cloud location and evolves over time. Therefore, the gravity-driven model developed by Atkinson [48] is adopted in this chapter to model the dispersion of dense gases, as follows:

$$R_t = \left(\frac{4}{3}\right)^{0.75} C_E^{0.5} \left(\frac{g'}{2\pi}\right)^{0.25} Q_v^{0.25} t^{0.75} \tag{3.9}$$

C_E represents an empirical constant and varies from 0.91 to 1.15. Q_v denotes the vapor cloud flow rate, g' denotes the relative density of the vapor. This approach was designed based on observations and simulation findings in low-wind conditions, typical of most major VCE accidents; therefore, it is more conservative to model vapor cloud dispersion when wind velocity is very low. (From the standpoint of risk management, it is more cautious because it will result in more destructive shock waves). Besides these analytical methods, software for dispersion modeling is also available such as ALOHA [49], PHAST [50], FLUENT [23], CFX [51], and FLACS [33]. Since a vast number of release scenarios are required in risk assessment, this chapter uses analytical approaches to model vapor cloud dispersion, overcoming the time-consuming feature of CFD tools.

3.3.3 *Step 3: Identification and Characterization of Ignition Sources*

To support the dynamic accident evolution assessment, the step of identification and characterization of ignition sources is critical. The first task of this step is to locate ignition sources, such as flares, boilers, and cars that can lead to immediate or delayed ignition. Measures for removing potential ignition causes are viewed as an important and realistic way to minimize the possibility of fire and explosion in the chemical and process industries; such measures can include reducing the flow rate during loading and unloading operations to avoid static electricity and ground rods to prevent lightning. In an industrial environment, however, it is difficult to remove all ignition sources. Furthermore, some significant vapors have been caused by ignition sources outside the manufacturing region [20, 30, 52]. As a result, this stage should define as many potential ignition sources as possible, both inside and outside the chemical plant. In other words, regardless of whether they are present or not, the potential ignition sources covered by the vapor cloud should be known (in the chemical plant or not). Under the presumption that the dangerous material inside the installation is fully released, the maximum vapor cloud can be measured using the dispersion model suggested in step 2.

Ignition is divided into immediate ignition and delayed ignition. The term "immediate ignition" refers to ignition that occurs at or near the release source and is fast enough to prevent the creation of a significant vapor cloud [38]. Thus, the probability of autoignition and the likelihood of static discharge, all of which can be regarded as unrelated to release time and vapor cloud dispersion, determine the likelihood of immediate ignition. The probability of autoignition (P_{Aut}) and the probability of static discharge (P_{Sta}) can be calculated [38], as follows:

$$P_{\text{Aut}} = 1 - 5000\text{e}^{-9.5(RT/AIT)} \tag{3.10}$$

$$P_{Sta} = 0.0024\left(RP/MIE^2\right)^{1/3} \tag{3.11}$$

AIT (°C) represents the autoignition temperature, MIE (J) represents the minimum ignition energy, RT (°C) denotes the actual release temperature, and RP (°C) represents the pressure of the released source. If $\frac{RT}{AIT} < 0.9$, $P_{Aut} = 0$ while $P_{Aut} = 1$ when $\frac{RT}{AIT} > 1.2$. This approach implies that if T is less than 200 °C and greater than the AIT, there is always a chance of non-immediate ignition. These correlations were created using a combination of ignition data and expert opinions. As a result, users should use these similarities with caution, choose conservative input parameter values, and do not read more precision into their forecasts than is realistic [38]. Other similarities and data will, of course, be conveniently incorporated into the existing approach based on the needs of particular users and applications.

A delayed ignition is any ignition other than an immediate ignition that has a delay period that allows for the creation and dispersion of a vapor cloud. Because of the dispersion of the vapor cloud, further ignition sources can become involved in the accident's evolution. The cumulative probability of ignition induced by an identified ignition source (IS) can be considered a function of time when the source is covered by the vapor cloud (t_{IS}) and the ignition effectiveness (ω), as shown in Eq. (4.9) [12].

$$P_{IS} = 1 - e^{-\omega t_{IS}} \tag{3.12}$$

The ignition effectiveness ω (s^{-1}) is determined by many factors such as ignition source types, ignition energy, and ignition control measures [53, 54]. The estimation of ω is a critical step for assessing the delayed ignition probability of a flammable vapor cloud. This chapter adopts the ignition probability estimation method developed by HSE [53]. The method considers the type of hazardous area, properties of on-site ignition sources (strength, frequency and duration of activity, and density), and ignition control measures in place. Because the ignition probability is calculated as an exponential time-independent function, the probability of ignition equals the probability of ignition in one minute, which can then be used to measure the ignition effectiveness. It should be remembered that this equation can only be used while the ignition source is active and the vapor cloud is present. As a result, before the ignition source is enabled or the vapor cloud arrives, the total likelihood should be equal to zero. Concerning the ignition caused by vehicles on a road or railway near the plant, the ignition probability can be calculated using the average traffic density d, as follows:

$$d = N_v L / v \tag{3.13}$$

N_v represents the number of vehicles per hour, L denotes the length of a road or railway section, v is the average velocity of the vehicle. Therefore, the probability of the ignition caused by vehicles on road or railway can be calculated as follows [12]:

$$P_{IS} = \begin{cases} d\left(1 - e^{-\omega t_{IS}}\right) d \leq 1 \\ 1 - e^{-d\omega t_{IS}}\ d > 1 \end{cases} \tag{3.14}$$

3.3.4 Step 4: Explosion Frequency and Delayed Time Assessment

In this step, a dynamic probabilistic tool is employed to assess the explosion frequency and model the temporal dependencies caused by a LOC event in chemical plants. There are multiple widely used dynamic probabilistic tools such as dynamic event tree [55], dynamic fault tree [56], dynamic bow-tie [57], dynamic Bayesian network [58], and Monte Carlo simulation [59]. Dynamic risk assessment methods can be divided into three categories: digraph-based methods,[2] explicit state-transition methods,[3] and implicit state-transition approaches[4] [60]. Dynamic event tree is a digraph-based method that is recommended for modeling system evolution. It can model stochastic behavior and possible dependencies among different events [60,

---

[2] A digraph-based method consists of a graph or diagram with nodes and directed arcs between the nodes, such as event tree and fault tree. The nodes indicate variables in the system or events impacting the system and the arcs indicate causal relationships between the nodes. Weights are also assigned to the arcs to indicate the relationships between nodes.

[3] Explicit state-transition methods can be regarded as generalized event trees, including, event sequence diagrams and explicit Markov state transition modeling. Different from digraph-based methods, explicit state-transition methods can provide a more literal representation of the plant state, the behavior of key process variables even operators, as well as hardware state changes.

[4] Implicit state transition methods can implicitly define the states via analyst-supplied rules, such as continuous event trees and dynamic logical analytical methods. To reduce computer memory storage requirements, these rules are used to form state-transition probabilities only when they are needed.

61]. In the current analysis, a discrete dynamic event tree (DDET) is used to explicitly present the accident evolution mechanism and potential accident scenarios. The DDET serves as a framework for simulating and analyzing the complex relationships between vapor cloud dispersion and ignition sources. Figure 3.4 is an example of DDET for accident evolution assessment.

The DDET shown in Fig. 3.4 represents a VCE evolution process with four ignition sources: autoignition, static discharge, ignition source 1, and ignition source 2. Using the bolded scenarios as an example, the likelihood of VCE/FF triggered by the ignition of source 1 at t_2 can be obtained as follows:

$$P_{t_2,IS1,VCE} = P_{LOC}P\left(\overline{I_{t_0}}|LOC\right)P\left(\overline{I_{t_1}}|\overline{I_{t_0}}\right)P\left(I_{t_2,IS1}|\overline{I_{t_1}}\right)P\left(VCE|I_{t_2,IS1}\right) \quad (3.15)$$

$$P_{t_2,IS1,FF} = P_{LOC}P\left(\overline{I_{t_0}}|LOC\right)P\left(\overline{I_{t_1}}|\overline{I_{t_0}}\right)P\left(I_{t_2,IS1}|\overline{I_{t_1}}\right)P\left(FF|I_{t_2,IS1}\right) \quad (3.16)$$

where $P_{t_2,IS1,VCE}$ represents the probability of VCE caused by ignition source 1 at t_2, $P_{t_2,IS1,FF}$ denotes the probability of FF caused by ignition source 1 at t_2, $P\left(VCE|I_{t_2,IS1}\right)$ represents the conditional probability of VCE given delayed ignition caused by ignition source 1 at t_2, $P\left(FF|I_{t_2,IS1}\right)$ denotes the conditional probability of FF given delayed ignition caused by ignition source 1 at t_2, $P\left(I_{t_2,IS1}|\overline{I_{t_1}}\right)$ denotes

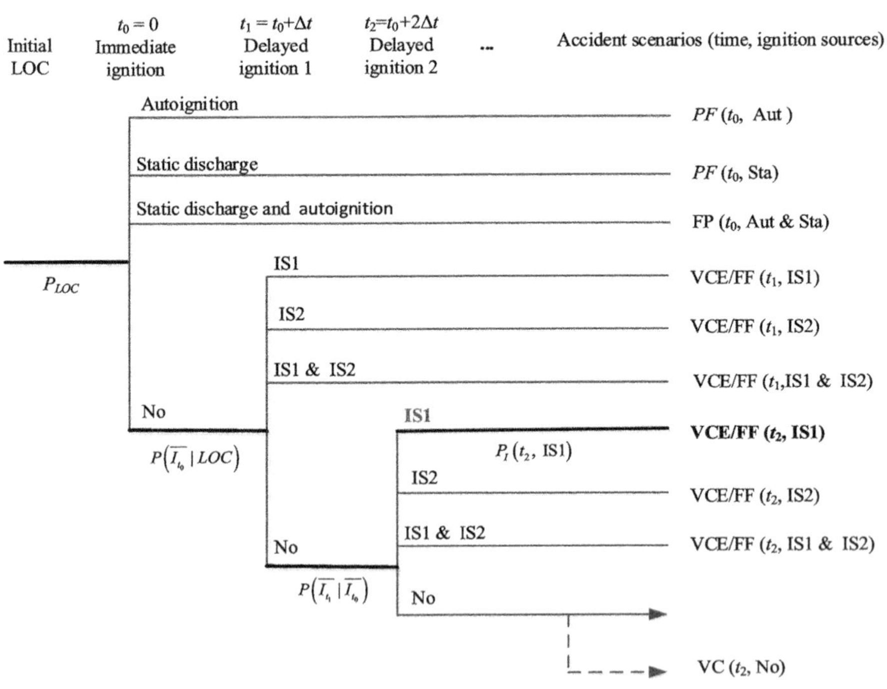

Fig. 3.4 A discrete dynamic event tree for accident evolution assessment (Chen et al. [1])

the conditional probability of the delayed ignition caused by ignition source 1 at t_2 given no ignition before time t_1, $P\left(\overline{I_{t1}}|\overline{I_{t0}}\right)$ represents the conditional probability of no ignition before time t_1 given no immediate ignition at time t_0, $P\left(\overline{I_{t_0}}|LOC\right)$ represents the probability of no immediate ignition at time t_0 given a LOC event, and P_{LOC} denotes the probability of the LOC event. $P\left(I_{t_2,IS1}|\overline{I_{t_1}}\right)$ can be calculated as follows:

$$P\left(I_{t_2,IS1}|\overline{I_{t_1}}\right) = \{P_{IS1}(t_2) - P_{IS1}(t_1)\} - \{P_{IS1}(t_2) - P_{IS1}(t_1)\}\{P_{IS2}(t_2) - P_{IS2}(t_1)\} \tag{3.17}$$

$P_{IS1}(t_2)$ represents the ignition probability of source 1 before time t_2, $P_{IS1}(t_1)$ denotes the ignition probability of source 1 before time t_1, $P_{IS2}(t_2)$ denotes the ignition probability of source 1 before time t_2, $P_{IS2}(t_1)$ represents the ignition probability of source 2 before time t_1. To simplify the probability calculation, a constant time step ($\Delta t = t_{i+1} - t_i$) is used in dynamic event tree analysis. The value of Δt can be chosen based on the necessary measurement accuracy and calculation time. The event tree can not terminate until the likelihood of the vapor cloud is less than a certain threshold, meaning that accident situations with a probability less than the threshold should be overlooked.

3.3.5 *Step 5: Overpressure Calculation*

This step introduces the Multi-Energy method to calculate the overpressure intensity of VCE scenarios identified in Sect. 3.3.4. According to the method, the intensity of overpressure depends on the scaled distance and the strength coefficient. The scaled distance (r_{sc}) can be obtained as follows [13]:

$$r_{sc} = \frac{r}{(E/P_a)^{1/3}} \tag{3.18}$$

$$E = M_f \times \Delta H \tag{3.19}$$

E represents the total combustion energy (J); P_a denotes the ambient pressure (Pa); r represents the distance from the center of the explosion (m); ΔH denotes the combustion heat of the flammable gas (J/kg). The strength coefficient represents the strength of the explosion blast, which depends on the obstacle density of the explosion area. The coefficient ranges from 1 to 10 and increases with increasing obstacle density. Several methods based on qualitative factors are available in the literature to estimate the strength coefficient, such as the method developed by Kinsella [62]. According to this method, The strength coefficient depends on three factors: (i) degree of obstruction of the area covered by the flammable gas, (ii) ignition energy, (iii) degree of confinement. The first factor is divided into three levels: high (obstacles in the gas cloud with a volume blockage fraction no less than 30% and with spacing between obstacles no more than 3 m), low (obstacles in a gas cloud with a blockage fraction less than 30% and/or spacing between obstacles over 3 m) and none (no obstacles within the flammable gas).

The ignition energy is divided into two types: high (high energy source) and low (low energy source). The factor of confinement is divided into two levels: confined (gas clouds, or parts of it, are confined by walls/barriers on two or three sides) and unconfined (gas cloud is not confined, other than by the ground). The strength coefficient can then be estimated using Table 3.2.

According to the scaled distance (r_{sc}) and the strength coefficient (SC) of the explosion, the scaled overpressure (P_{sc}) can be read from a blast chart, as follows:

The blast chart was obtained by numerical simulations using a Flux-Corrected Transport code [63]. In Fig. 3.5, the horizontal axis represents the scaled distance (r_{sc}), the inner vertical axis represents the strength coefficient (SC) while the outside vertical axis represents the scaled overpressure (P_{sc}). As a result, the overpressure can be obtained, as follows:

$$P_o = P_{sc} \times P_a \tag{3.20}$$

P_o represents the intensity of overpressure. It should be marked that uncertainty exists in all the widely used calculation methods. For instance, the fraction of energy released as shock wave is the main uncertainty parameter in the Equivalent TNT, while that is the strength coefficient in the Multi-Energy method.

Table 3.2 Blast Strength Index (Chen et al. [1])

Ignition energy		Degree of obstruction			Confinement		Strength coefficient
Low (L)	High (H)	High (H)	Low (L)	No (N)	Confined (C)	Unconfined (U)	
	H	H			C		7–10
	H	H				U	7–10
L		H			C		5–7
	H		L		C		5–7
	H		L			U	4–6
	H			N	C		4–6
L		H				U	4–5
	H			N		U	4–5
L			L		C		3–5
L			L			U	2–3
L				N	C		1–2
L				N		U	1

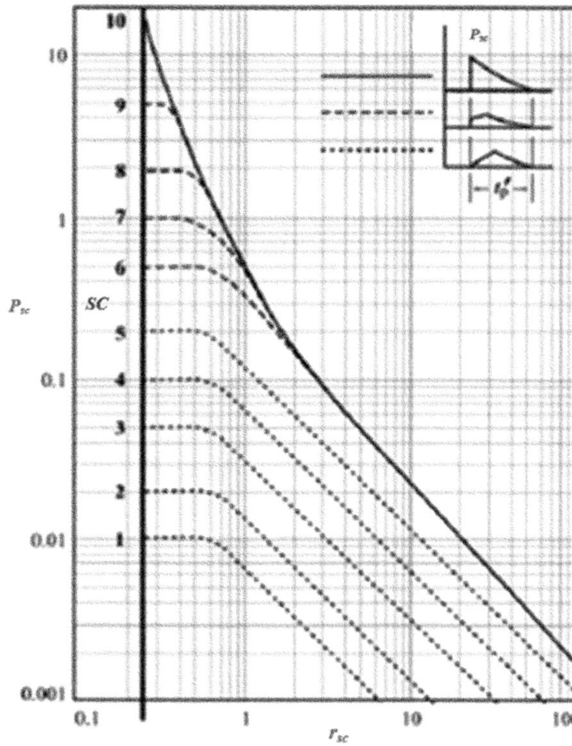

Fig. 3.5 The blast chart used in the multi-energy method (Van Den Bosh and Weterings [18])

3.3.6 Step 6: Escalation Assessment

To address the uncertainty in the escalation induced by a VCE, probability models are always used to assess the vulnerability of installations exposed to VCEs. For instance, Bagster and Pitblado [64] proposed a probabilistic approach by defining a damage probability function based on safety distances. Besides, Khan and Abbasi [65] used a probit function to calculate the failure probability caused by overpressure. The probit function was firstly developed by Eisenberg et al. [66] and only peak overpressure was considered in this chapter, as follows:

$$Y = c_5 + c_6 \ln(P_o) \tag{3.21}$$

Y represents the probit value; c_5 and c_6 are constants. For atmospheric tanks, c_5 and c_6 are equal to −9.36 and 1.43, respectively [67]. Then the damage probability P_r can be calculated using the cumulative standard normal distribution, the same as the distribution function used in Chap. 2.

According to the methodology, the escalation probability caused by VCEs and the damage probabilities of installations exposed to overpressure can be obtained.

3.4 Case Study

The Buncefield oil depot [27] is used as a case study to illustrate the dynamic approach proposed in this study. It has a total storage capacity of 273,000 m^3. The main 35 storage tanks in the plant are numbered and shown in Fig. 3.6.

According to the methodology illustrated in Sect. 3.3, hazardous installations should be identified and possible LOC scenarios need to be characterized. The main facilities of the plant are 35 ground storage tanks, pipes linking with these installations, loading and unloading facilities, and other components such as valves and pumps. Possible LOC scenarios are releases from storage tanks, tank overfilling during loading and unloading operations, and leakages from pipelines. To illustrate the proposed methodology, an overfilling scenario is considered based on a past accident scenario [27]. The mass flow rate is 108 kg/s [47]. The probability of the overfilling scenario is 5×10^{-2} per year [68].

Based on the characteristics of the LOC scenario, a vapor cloud dispersion analysis has to be carried out in step 2. The ratio of the evaporation rate to flow rate α is approximately equal to 0.18 and the vapor addition rate to the cloud is estimated as 199 m^3/s [47]. Since the gasoline vapor is denser than air, the gravity-driven model is adopted, considering $C_E = 1$ and $g' = 0.5$ [48]. As a result, the vapor cloud radius R at each time slice can be calculated according to Eq. (3.9). The autoignition probability P_{ia} is considered to be zero because the ambient temperature is very low. According to Eq. (3.11), the ignition probability of static discharge P_{is} is obtained: 0.0156. According to the accident investigation [69, 70], two possible ignition sources were identified in the oil storage depot: the pump house (IS1) and the Northgate emergency generator (IS2). Both the primary ignition probabilities of the two sources are considered to be 0.1/min [53]. Consequently, the parameter of ω in Eq. (3.12) is equal to 0.0018. IS1 is active after $t = 1.5$ min while IS2 is active at $t = 5$ min according to the results of the dispersion model.

Fig. 3.6 The layout of the Buncefield (UK) oil depot before 2005 (Chen et al. [1])

The next step is to estimate the explosion frequency and possible delayed time using a discrete dynamic event tree with a time step $\Delta t = 1$ min. Figure 3.7 shows the dynamic event tree up until 25 min with 66 possible scenarios.

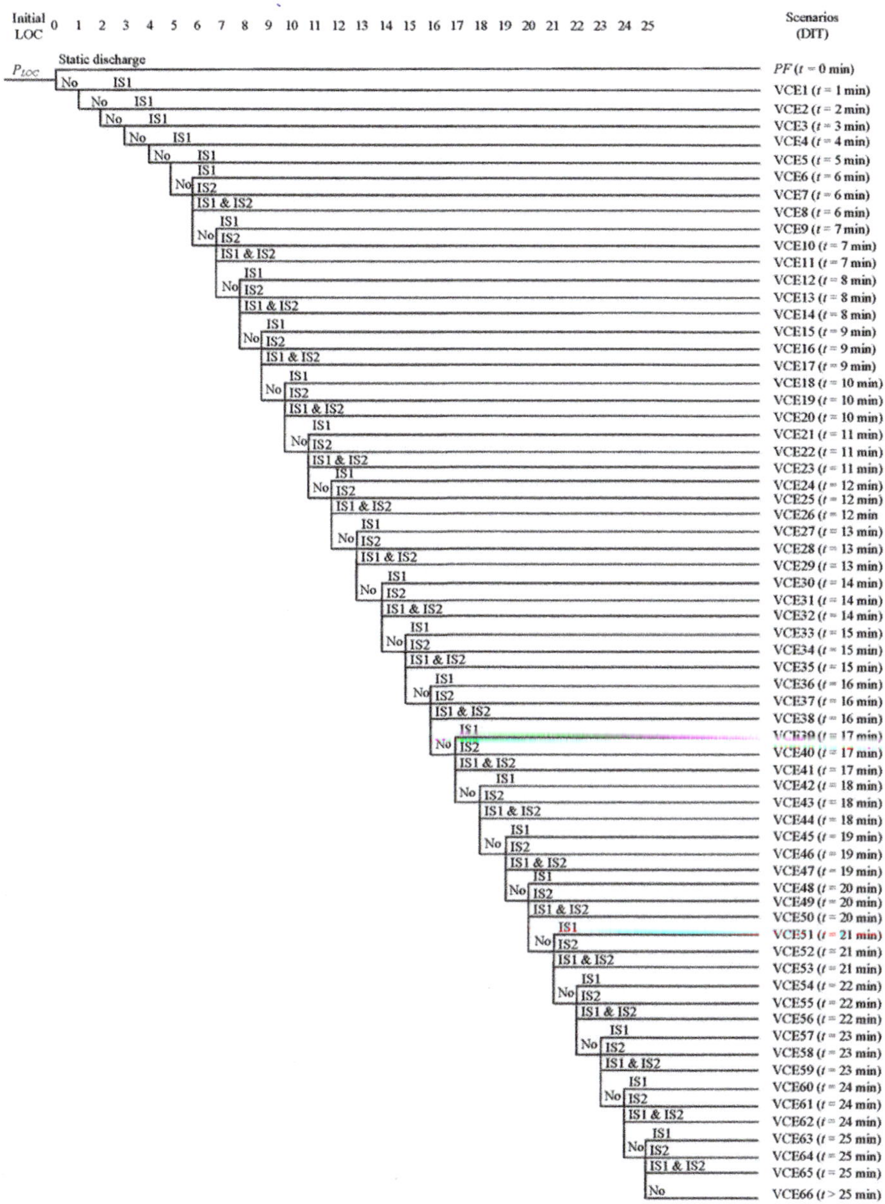

Fig. 3.7 A discrete dynamic event tree for VCE assessment (Chen et al. [1])

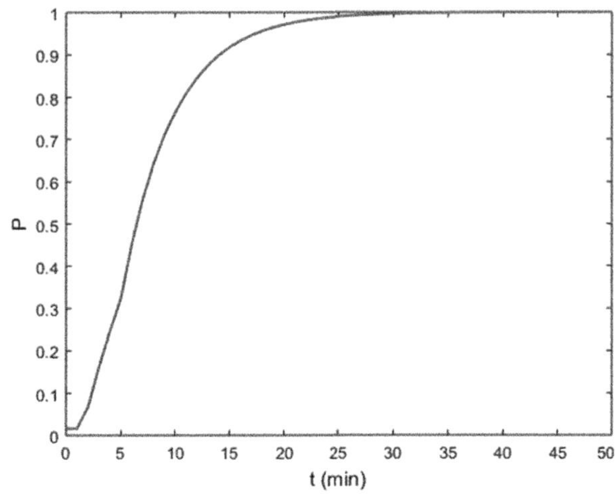

Fig. 3.8 The cumulative explosion probability over time (Chen et al. [1])

Based on the event tree, the cumulative explosion probability can be obtained, as shown in Fig. 3.8. The cumulative probability increases over time while the increase rate decreases over time when the second ignition source is covered by flammable gas. Finally, the cumulative explosion approaches 1.

Based on the spatial–temporal dispersion results, the overpressure caused by VCE at each discrete time can be obtained by using the Multi-Energy method. To be conservative, the strength coefficient is considered to be 10 for all the VCE scenarios. The overpressures caused by different explosions are shown in Fig. 3.9. The overpressure increases over DIT due to the increase of total explosion energy. The maximum overpressure can be up to 2000 kPa which is consistent with the Buncefield explosion investigations [11, 19]. But the overpressure rapidly decreases with increasing the distance from the center of the explosion. So the nearby equipment may be more vulnerable. Figure 3.10 shows the damage probability of tanks caused by possible VCE scenarios. Tanks 25–35 have a lower damage probability than other tanks since they are situated at a larger distance from the release tank. The number of expected damaged tanks (the sum of damage probability of all the 35 tanks) is 16. In that case, domino effects may be inevitable due to possible synergistic effects.

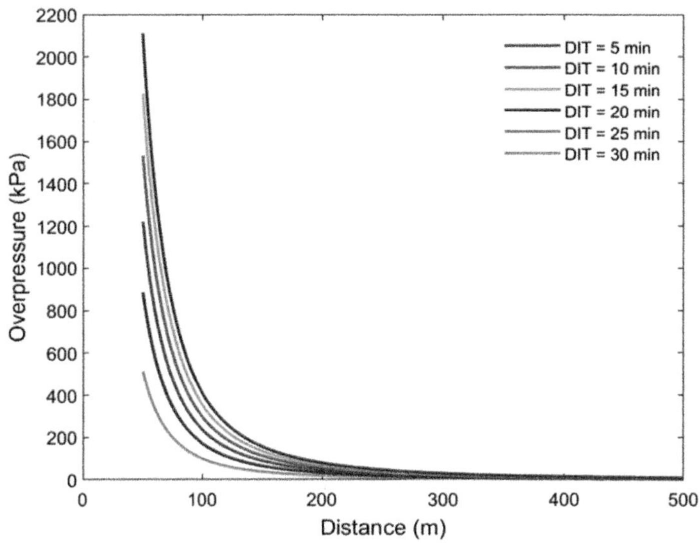

Fig. 3.9 Calculation results of overpressure (Chen et al. [1])

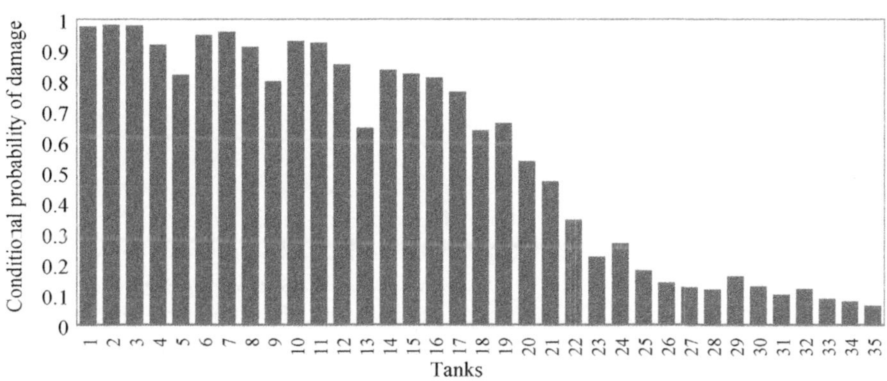

Fig. 3.10 Conditional probability of damage given an overfilling at tank 2 (Chen et al. [1])

3.5 Discussion

The case study results demonstrate that the developed dynamic assessment methodology can model the influence of the spatial–temporal evolution of a vapor cloud and the uncertainty of delayed ignition on the vulnerability of installations subject to VCEs. This section discusses some essential issues related to this research.

In this study, the Multi-Energy method was adopted to obtain the overpressure caused by VCEs. The most difficult issue in the application of this method is to determine the explosion strength coefficient based on the obstacle density of process plants. Although TNO published the yellow book to guide the application of the Multi-Energy method, it is still difficult for users to determine the value of the strength coefficient. However, the value of the explosion strength coefficient plays an important role in overpressure estimation. For instance, the maximum conditional damage probability subject to VCEs caused by overfilling at T2 decreases to 0.35 if the strength coefficient of 10 is substituted by 3. Therefore, the strength coefficient should be determined by meticulously analyzing the layout of a process. Otherwise, the worst-case scenario (with a strength coefficient equal to 10) may be adapted to be conservative.

This study adopted an analytic method to predict the dispersion of vapor cloud, neglecting the VCE dilution with distance. To address vapor cloud dilution, upper and lower explosion limits, CFD methods may be integrated into this methodology to obtain more accurate dispersion results in future studies. With the rapid development of computational techniques, applying CFD methods in risk assessment may become easier and acceptable in the future.

Monte Carlo simulation can also be integrated into this methodology to evaluate the frequency of each scenario when the number of possible accident scenarios becomes too large due to the increase in the number of ignition sources. This study neglects second and higher-level escalations which could be considered in the future. Besides, by the application of Bayesian theory, this study may be extended to identify the most likely ignition sources based on evidence-based reasoning.

3.6 Conclusions

This chapter provides a new methodology based on a dynamic event tree to model the escalation triggered by VCEs. Unlike previous work on risk assessment of VCEs, the developed methodology considers the time dependencies in vapor cloud dispersion and the uncertainty in delayed ignition. In that case, the explosion risk changes over time, and the characteristics of possible large VCEs can be reflected to avoid underestimation of their risks. According to the methodology, the vulnerability of installations to VCEs is related to the congestion of the plant layout and DIT, and a long-delayed explosion may result in more severe consequences. The delayed ignition time depends on the distance between the release position and the ignition sources, the type of ignition sources, and the ignition control measures in place. The ignition control measures can delay (if not fully eliminate) the VCE while a delayed ignition (which might be considered a good thing in some circumstances) could lead to a severe explosion. This work shows how DET can effectively be used to assess domino effects caused by VCEs, addressing severe consequences caused by large VCEs.

References

1. Chen C, Khakzad N, Reniers G (2020) Dynamic vulnerability assessment of process plants with respect to vapor cloud explosions. Reliab Eng Syst Saf 200
2. Chen C, Reniers G, Khakzad N (2019) Exploring loading and unloading operations in relation to domino effects in chemical industrial parks. Chem Eng Trans 75
3. Chen C (2021) A dynamic and integrated approach for modeling and managing domino-effects. Delft University of Technology
4. Lees F (2012) Lees' loss prevention in the process industries: hazard identification, assessment and control. Butterworth-Heinemann
5. Pietersen C (1990) Consequences of accidental releases of hazardous material. J Loss Prev Process Ind 3(1):136–141
6. Chen C, Reniers G, Khakzad N (2020) A thorough classification and discussion of approaches for modeling and managing domino effects in the process industries. Saf Sci 125
7. Khan FI, Abbasi S (1999) Major accidents in process industries and an analysis of causes and consequences. J Loss Prev Process Ind 12(5):361–378
8. Bellamy L, Geyer T, Astley J (1989) Evaluation of the human contribution to pipework and in-line equipment failure frequencies. Health and Safety Executive
9. Abdolhamidzadeh B, Abbasi T, Rashtchian D, Abbasi SA (2011) Domino effect in process-industry accidents—an inventory of past events and identification of some patterns. J Loss Prev Process Ind 24(5):575–593
10. Taylor JR (2003) Risk analysis for process plant, pipelines and transport. Routledge
11. Taveau J (2012) The Buncefield explosion: were the resulting overpressures really unforeseeable? Process Saf Prog 31(1):55–71
12. Uijt de Haag A (1999) Guidelines for quantitative risk assessment. Committee for the Prevention of Disasters, The Hague, NL
13. Assael MJ, Kakosimos KE (2010) Fires, explosions, and toxic gas dispersions: effects calculation and risk analysis. CRC Press
14. Van den Berg A, Lannoy A (1993) Methods for vapour cloud explosion blast modelling. J Hazard Mater 34(2):151–171
15. Lea C, Ledin H (2002) A review of the state-of-the-art in gas explosion modelling. Health and Safety Laboratory Buxton
16. Cozzani V, Salzano E (2004) The quantitative assessment of domino effects caused by overpressure: part I. Probit models. J Hazard Mater 107(3):67–80
17. Baker QA, Tang MJ, Scheier EA, Silva GJ (1996) Vapor cloud explosion analysis. Process Saf Prog 15(2):106–109
18. Van Den Bosh C, Weterings R (1997) Methods for the calculation of physical effects (Yellow Book). Committee for the Prevention of Disasters, The Hague, NL
19. Mishra KB, Wehrstedt K-D, Krebs H (2013) Lessons learned from recent fuel storage fires. Fuel Process Technol 107:166–172
20. Mishra KB, Wehrstedt K-D, Krebs H (2014) Amuay refinery disaster: the aftermaths and challenges ahead. Fuel Process Technol 119:198–203
21. Baker QA, Doolittle CM, Fitzgerald GA, Tang MJ (1998) Recent developments in the Baker-Strehlow VCE analysis methodology. Process Saf Prog 17(4):297–301
22. Qiao A, Zhang S (2010) Advanced CFD modeling on vapor dispersion and vapor cloud explosion. J Loss Prev Process Ind 23(6):843–848
23. Tauseef SM, Rashtchian D, Abbasi T, Abbasi SA (2011) A method for simulation of vapour cloud explosions based on computational fluid dynamics (CFD). J Loss Prev Process Ind 24(5):638–647
24. Sierra D, Briceno J, Buitrago H, Rozo B, Montecchi L, Mura I (2018) Probabilistic modeling of failure domino effects in chemical plants
25. Ramírez-Marengo C, Diaz-Ovalle C, Vázquez-Román R, Mannan MS (2015) A stochastic approach for risk analysis in vapor cloud explosion. J Loss Prev Process Ind 35:249–256

26. Ouddai R, Chabane H, Boughaba A, Frah M (2012) The Skikda LNG accident: losses, lessons learned and safety climate assessment 35(6):518–533
27. Buncefield Major Incident Investigation Board (2008) The Buncefield incident, 11 December 2005: the final report of the Major Incident Investigation Board. Health and Safety Executive
28. CSB (2015) Final investigation report Caribbean petroleum tank terminal explosion and multiple tank fires. U.S. Chemical Safety and Hazard Investigation Board
29. Siddiqui N, Tauseef S, Abbasi S, Rangwala AS (2018) Advances in fire and process safety. Springer, Berlin
30. Ministry of Emergency Management of the People's Republic of China (2018) The report of the State Council Work Safety Committee on the "11·28" major explosion-fire accident in Zhangjiakou of Hebei Province. http://www.chinasafety.gov.cn/gk/xxgk/201811/t20181130_222839.shtml. Accessed 22 Dec 2018
31. Johnson D, Tam V (2017) Why DDT is the only way to explain some vapor cloud explosions. Process Saf Prog 36(3):292–300
32. Atkinson G, Hall J, McGillivray A (2017) Review of vapour cloud explosion incidents. Health and Safety Executive, Buxton
33. Dasgotra A, Varun Teja GVV, Sharma A, Mishra KB (2018) CFD modeling of large-scale flammable cloud dispersion using FLACS. J Loss Prev Process Ind 56:531–536
34. Maremonti M, Russo G, Salzano E, Tufano V (1999) Post-accident analysis of vapour cloud explosions in fuel storage areas. Process Saf Environ Prot 77(6):360–365
35. Gant SE, Atkinson GT (2011) Dispersion of the vapour cloud in the Buncefield Incident. Process Saf Environ Prot 89(6):391–403
36. Badri N, Nourai F, Rashtchian D (2013) A multivariable approach for estimation of vapor cloud explosion frequencies for independent congested spaces to be used in occupied building risk assessment. Process Saf Environ Prot 91(1–2):19–30
37. Khakzad N, Reniers G, Abbassi R, Khan F (2016) Vulnerability analysis of process plants subject to domino effects. Reliab Eng Syst Saf 154:127–136
38. Moosemiller M (2011) Development of algorithms for predicting ignition probabilities and explosion frequencies. J Loss Prev Process Ind 24(3):259–265
39. Alileche N, Olivier D, Estel L, Cozzani V (2017) Analysis of domino effect in the process industry using the event tree method. Saf Sci 97:10–19
40. Atkinson G, Cowpe E, Halliday J, Painter D (2017) A review of very large vapour cloud explosions: cloud formation and explosion severity. J Loss Prev Process Ind 48:367–375
41. Cozzani V, Tugnoli A, Salzano E (2009) The development of an inherent safety approach to the prevention of domino accidents. Accid Anal Prev 41(6):1216–1227
42. Villa V, Reniers GLL, Paltrinieri N, Cozzani V (2017) Development of an economic model for the allocation of preventive security measures against environmental and ecological terrorism in chemical facilities. Process Saf Environ Prot 109:311–339
43. Khakzad N, Martinez IS, Kwon H-M, Stewart C, Perera R, Reniers G (2018) Security risk assessment and management in chemical plants: Challenges and new trends. Process Saf Prog 37(2):211–220
44. Baybutt P (2017) Issues for security risk assessment in the process industries. J Loss Prev Process Ind 49:509–518
45. Reniers G, Khakzad N, Gelder PV (2018) Security risk assessment in the chemical and process industry. Walter de Gruyter GmbH & Co KG
46. CCPS (1996) Guidelines for use of vapor cloud dispersion models. Wiley-AIChE
47. Atkinson G, Coldrick S (2012) Vapour cloud formation: experiments and modelling. Health and Safety Laboratory, Debyshire
48. Atkinson G (2017) Development of heavy vapour clouds in very low wind speeds. J Loss Prev Process Ind 48:162–172
49. Tseng J, Su T, Kuo C (2012) Consequence evaluation of toxic chemical releases by ALOHA. Procedia Eng 45:384–389
50. Zarei E, Jafari MJ, Badri N (2013) Risk assessment of vapor cloud explosions in a hydrogen production facility with consequence modeling. J Res Health Sci 13(2):181–187

51. Qi R, Ng D, Cormier BR, Mannan MS (2010) Numerical simulations of LNG vapor dispersion in Brayton Fire Training Field tests with ANSYS CFX. J Hazard Mater 183(1–3):51–61
52. Dweck J, Boutillon S, Asbill S (2004) Deadly LNG incident holds key lessons for developers, regulators. Pipeline Gas J 39–42
53. Rew P, Daycock L (2004) Development of a method for the determination of on-site ignition probabilities, HSE contractor report WSA/226. HSE Books, London, UK
54. Šrekl J, Golob J (2011) New approach to calculate the probability of ignition. J Loss Prev Process Ind 24(3):288–291
55. Acosta C, Siu N (1993) Dynamic event trees in accident sequence analysis: application to steam generator tube rupture. Reliab Eng Syst Saf 41(2):135–154
56. Dugan JB, Bavuso SJ, Boyd MA (1992) Dynamic fault-tree models for fault-tolerant computer systems. IEEE Trans Reliab 41(3):363–377
57. Khakzad N, Khan F, Amyotte P (2012) Dynamic risk analysis using bow-tie approach. Reliab Eng Syst Saf 104:36–44
58. Khakzad N (2015) Application of dynamic Bayesian network to risk analysis of domino effects in chemical infrastructures. Reliab Eng Syst Saf 138:263–272
59. Durga Rao K, Gopika V, Sanyasi Rao VVS, Kushwaha HS, Verma AK, Srividya A (2009) Dynamic fault tree analysis using Monte Carlo simulation in probabilistic safety assessment. Reliab Eng Syst Saf 94(4):872–883
60. Siu N (1994) Risk assessment for dynamic systems: an overview. Reliab Eng Syst Saf 43(1):43–73
61. Aldemir T (2018) Advanced concepts in nuclear energy risk assessment and management. World Scientific Publishing Company Pte. Limited
62. Kinsella K (1993) A rapid assessment methodology for the prediction of vapour cloud explosion overpressure. In: International conference and exhibition on safety, health and loss prevention in the oil, chemical and process industries, Singapore
63. van den Berg A (1980) A ID variable flame speed blast simulation code using a Flux-Corrected Transport algorithm. Report No PML 162
64. Bagster DF, Pitblado RM (1991) The estimation of domino incident frequencies—an approach. Process Saf Environ Prot 69(4):195–199
65. Khan FI, Abbasi S (1998) Models for domino effect analysis in chemical process industries. Process Saf Prog 17(2):107–123
66. Eisenberg NA, Lynch CJ, Breeding RJ (1975) Vulnerability model. A simulation system for assessing damage resulting from marine spills. Enviro Control Inc Rockville Md
67. Zhang M, Jiang J (2008) An improved probit method for assessment of domino effect to chemical process equipment caused by overpressure. J Hazard Mater 158(2–3):280–286
68. Hauptmanns U (2004) Semi-quantitative fault tree analysis for process plant safety using frequency and probability ranges. J Loss Prev Process Ind 17(5):339–345
69. Bakke JR, van Wingerden K, Hoorelbeke P, Brewerton B (2010) A study on the effect of trees on gas explosions. J Loss Prev Process Ind 23(6):878–884
70. Health and Safety Executive (2009) Buncefield explosion mechanism phase 1. Health and Safety Executive

Chapter 4
Risk Assessment of Coupled Hazardous Scenarios

4.1 Introduction

In the process and chemical industry,[1] major accident scenarios such as fire, explosion, and toxic gas dispersion may occur due to the loss of containment of hazardous materials [3–6]. Among these major accident scenarios, fire is the most common scenario (44%), followed by explosion (36%). 20% of all major accidents are only toxic dispersion without fire and explosion, while 30% of the explosion and fire scenarios involve toxic substances [7]. Since chemical industrial areas are usually congested with many hazardous installations, a fire or explosion may lead to domino effects, resulting in multiple accident scenarios. As a result, major accident scenarios such as fire, explosion, and toxic release can simultaneously or sequentially be present in one disaster due to the evolution of domino effects.

In light of past disasters listed in Chap. 3 and since unpredicted hazards may result in severe consequences, modeling the spatial–temporal evolution of hazards and events originating from a release of hazardous materials in industrial areas is essential for protecting staff, residents, and emergency rescuers. For example, in the Tianjin port disaster in 2015, caused by the spontaneous ignition of nitrocellulose, many of the emergency rescuers were killed in the disaster due to an unpredicted evolution of the fire to an explosion. To avoid such catastrophic disasters, many post-accident analyses were conducted on the prediction of overpressure induced by explosions [8–12] and vapor cloud dispersion [13–15]. Besides, a lot of work has been done on vulnerability assessment of installations to VCEs, risk assessment of domino effects caused by VCE [16–21] and domino effects triggered by fire [22–24]. In terms of the evolution of fire, the time to failure of equipment subject to fire is critical for the vulnerability of installations [25]. As a result, dynamic methods were used to assess the vulnerability of installations exposed to fire and fire-induced domino effects [26–29]. Among these dynamic tools, both the dynamic graph approach and

[1] This chapter is mainly based on two publications: Chen et al. [1] and Chen [2].

C. Chen et al., *Integrating Safety and Security Management to Protect Chemical Industrial Areas from Domino Effects*, Springer Series in Reliability Engineering,
https://doi.org/10.1007/978-3-030-88911-1_4

the dynamic Bayesian network approach can model the spatial–temporal evolution of domino effects caused by fire and visualize the escalation paths of fire [3, 26]. Besides, Monte Carlo simulation was used to address the evolution uncertainty of domino effects [30–32].

In view of the research attempts on VCE hazard or fire hazard evolution, little attention has been paid to dynamic hazard evolution of possible toxic release, VCE and fire in a catastrophic disaster, and the assessment of human exposure to multiple hazards sequentially. This chapter, therefore, aims to establish a dynamic method for human and facility vulnerability assessment considering the spatial–temporal evolution of multiple hazards such as toxic release, VCE, and fire. Both the uncertainty of ignition and the uncertainty of the evolution of different hazards are considered in the present study. The dynamic graph model is developed in Sect. 4.2. The graph update rules and a simulation algorithm based on dynamic Monte Carlo simulation are introduced in Sect. 4.3. The conclusions drawn from this chapter are summarized in Sect. 4.4.

4.2 Modeling

Agent-based modeling (ABM) is a powerful modeling technique that has been widely used in many domains in recent decades, such as complex adaptive systems, biology, business [33]. Compared with other modeling tools based on events or processes, ABM focuses on agent behaviors (attributes and interactions), and it is very flexible to model socio-technical systems [30, 34, 35]. Graph-based methods can provide a visible structure (graph or network) to represent the complex agent interactions. Consequently, graph-based approaches are often used to model interacting agents in a system [36, 37]. By applying graphs, the agents are represented by nodes, and their dependencies are denoted by edges [23, 38, 39]. According to this principle, a chemical plant can be regarded as a system (a graph), while the hazardous installations within the plant can be modeled as agents (nodes). In that case, graph metrics such as in-closeness and out-closeness can be used to assess the domino effect potential and the damage likelihood of installations being damaged by escalation effects [23, 40]. The spatial–temporal evolution characteristics of fire-induced escalations can then be addressed by dynamic graphs [27].

In terms of modeling the evolution between different accident scenarios, it is difficult to address the uncertainty in scenario evolution by merely using the dynamic graph approach. To deal with the uncertainty, the Monte Carlo method can be integrated with the dynamic graph approach. Monte Carlo simulation is a commonly used tool to model uncertainties and avoid complex mathematical calculations [41–43]. The new integrated approach based on dynamic graphs and Monte Carlo simulation is called Dynamic Graph Monte Carlo (DGMC). The DGMC is an ABM method consisting of a dynamic graph with random parameters and a Monte Carlo simulation for modeling uncertainties.

According to the developed DGMC method, we can model the evolution of multiple accident scenarios in coupled domino effect events and dynamically assess human vulnerability exposed to these scenarios. The DGMC is defined as a dynamic graph with a nine-tuple, as follows:

$$DGMC = (M, N, K, C, O, Q, T) \tag{4.1}$$

4.2.1 Graph Nodes

In the DGMC model, graph nodes are used to model installations (N), human positions (M), and ignition sources (K).

(1) **Hazardous installations**

n represents the numbering of hazardous installations, and a total of N installations are assumed in a chemical plant. According to possible major accidents in industrial areas and the vulnerability characteristics of hazardous installations, the states of hazardous installations are divided into five types, as shown in Table 4.1.

The state "operational" is an initial state indicating that the installation is undamaged and operational. The state "extinguished" is a termination state in which the installation is damaged without any hazardous effects. Unstable states of "release", "fire" and "VCE" represent major accident scenarios of the loss of containment of hazardous materials, fire, and vapor cloud explosion, respectively. These unstable states can be harmful to installations and humans. The state of "fire" is caused by an immediate ignition, while the state of "VCE" results from delayed ignition. Table 4.2 shows human states, including one initial state "safe" and two terminal states "injured" and "dead".

Table 4.1 States of hazardous installations (Chen et al. [1])

State	Description
Operational	The hazardous installation is not physically damaged and is operational
Release	The hazardous installation is physically damaged, resulting in the loss of containment of hazardous materials and/or poisoning humans nearby
Fire	The installation is on fire due to immediate ignition, causing heat radiation on humans and/or other installations
VCE	The installation's loss of containment induces a vapor cloud explosion due to delayed ignition
Extinguished	The installation is physically damaged but does not generate any hazardous effects

Table 4.2 Human states (Chen et al. [1])

State	Description
Safe	The human does not receive any hazardous effects
Injured	The human is injured due to exposure to toxic gas, heat radiation, or overpressure
Dead	The human is decreased due to exposure to toxic gas, heat radiation, or overpressure

Table 4.3 States of ignition sources (Chen et al. [1])

State	Description
Inactive	Flammable vapor is not present at the ignition source, or the vapor concentration is out of the flammability limit
Active	Flammable materials are present at the ignition source, and the concentration of the vapor is between the lower and upper flammability limits
Ignited	The ignition source has ignited the flammable vapor

(2) **Human positions**

m represents the numbering of human positions in which humans may suffer from toxicity from toxic gases, heat radiation caused by fire, or overpressure induced by VCEs. A total of M human positions are assumed in the chemical plants. People at different places and times may suffer different hazardous effects caused by major accident scenarios. According to possible hazardous effects and human vulnerability characteristics in a chemical plant, the states of humans are divided into three types, as shown in Table 4.2.

According to Table 4.2, a person in a chemical plant may either (i) be safe without any hazardous effects or (ii) may be injured or (iii) may be dead due to hazardous effects.

(3) **Ignition sources**

k represents the numbering of ignition sources that may ignite flammable gases. According to ignition requirements, the states of ignition sources can be categorized into three types: "inactive", "active" and "ignited", as shown in Table 4.3.

As shown in Table 4.3, the states of ignition include one initial state (inactive), one terminal state (ignited), and one intermediate state (active). It should be noted that all the preceding states are time-dependent and may be updated with the spatial–temporal evolution of the hazards.

4.2.2 Graph Edges

In the DGMC model, directed edges represent the hazardous effects that may cause damage to hazardous installations or be harmful to humans. In this chapter, we consider three types of hazardous effects: the heat radiation induced by fire, the overpressure caused by VCEs, and the toxicity induced by a toxic cloud. Accordingly, we defined four types of directed edges: the ignition effects from released installations to ignition sources, the toxic effects from installation nodes to human nodes, the heat radiation from installation nodes to installation nodes or human nodes, and the overpressure effects from installation nodes to installation nodes or human nodes.

(1) **Delayed ignition**

If the hazardous materials which have been released are ignited immediately, no directed edges need to be established between release installations and ignition sources. Otherwise, when a flammable gas covers an ignition source, a directed edge from the installation where the material was released towards the ignition source should be established. If the flammable gas is ignited subsequently, all the directed edges from the released installations where materials have been released to ignition sources should be removed.

(2) **Acute intoxication**

If a human position is covered by toxic gas, a directed edge should link the released installation where the toxic material was released and the person's position. The acute intoxication of humans at the position covered by a toxic gas is determined by the toxic concentration (C_t) and exposure time (t_e). In this chapter, the probit function of acute intoxication [44] is used to obtain the death probability caused by human exposure to toxic gas, as follows:

$$Y_t = c_7 + c_8 \ln\left(C_t^{c_9} \times t_e\right) \tag{4.2}$$

Y_t represents the probit value of human vulnerability exposure to toxic gas. c_7, c_8, and c_9 represent the parameters of the probit function that change with different toxic substances. These parameters for some common toxic materials can be obtained from Van Den Bosh et al. [44]. Besides the toxic concentration and exposure time, other factors such as demographics (e.g., ages) and Personal Protection Equipment (PPE) may partially affect the death induced by toxic gases. For instance, if an employee wears a gas mask, the likelihood of death caused by toxic gas can be significantly reduced due to the filtering effect induced by the mask.

(3) **Damage induced by VCEs**

Since overpressure can damage installations and be harmful to humans, directed edges should be established from the installation where the overpressure generates to both installations and human positons. In the DGMC model, O represents the directed edge caused by overpressure. As illustrated in Chap. 3, several methods for overpressure calculation are available, such as the TNT equivalent method, the Baker-Strehlow method, and CFD simulation. The TNT equivalent method is a simple method based on the TNT explosion mechanism, neglecting the effects of space configuration, ignition sources, and flammable gas distribution. As a result, applying

Table 4.4 Probit function parameters of overpressure (Chen et al. [1])

Installations	Atmospheric	Pressurized	Elongated	Auxiliary	Human
c_5	−9.36	−14.44	−12.22	−12.42	−77.1
c_6	1.43	1.82	1.65	1.64	6.91

the TNT equivalent method in VCE estimation may underestimate the overpressure. The Multi-Energy method is developed for gas explosions by dividing an explosion into a number of sub-explosions and considering the roles of congestion levels, ignition, and gas distribution in obstructed areas. For quantitative analysis, the Netherlands Organization for Applied Scientific Research (TNO) recommends the Multi-Energy method [45]. Therefore, the Multi-Energy method [46] is adopted in this chapter for calculating the overpressure intensity received by different installations and humans. More details related to this method are described in Chap. 3.

The damage probability of hazardous installations and the death probability induced by overpressure can also be calculated by applying probit functions. The damage probability of installations and the death probability of humans depend on the overpressure intensity and the vulnerability characteristics of installations and humans exposed to overpressure, as follows:

$$Y_p = c_5 + c_6 \ln(P_o) \tag{4.3}$$

Y_p denotes the probit value with respect to overpressure generated by VCEs. c_5 and c_6 represent parameters related to the vulnerability characteristics of installations and humans, as shown in Table 4.4.

(4) **Damage induced by fires**

The heat radiation induced by fire can be harmful to humans and installations. If an installation is on fire, directed edges should be added from the installation to other installations and humans. Q is a matrix representing the heat radiations generated by installations in "fire" states. $q_{i,j}$ is an element of the matrix representing the heat radiation induced by an installation i in a "fire" state to an installation or a human j, as follows:

$$Q = \begin{bmatrix} 0 & q_{1,2} & \cdots & q_{1,m+n} \\ q_{2,1} & 0 & \cdots & q_{2,m+n} \\ \cdots & q_{i,j} & 0 & \cdots \\ q_{m,1} & \cdots & q_{m,m+n-1} & q_{m,m+n} \end{bmatrix} \tag{4.4}$$

$q_{i,j}$ can be calculated by the application of ALOHA. It should be remarked that Q is not a square matrix in this chapter since humans can only receive but not induce heat radiation. Considering possible synergistic effects [3] caused by multiple installations in the "fire" states, the heat radiation received by an installation or human j (Q_j) can be obtained, as follows:

$$Q_j = \sum_{i=1}^{m} q_{i,j} \tag{4.5}$$

According to the heat radiation matrix, the Domino Evolution Graph (DEG) model developed in Chap. 2 can be applied to calculate the evolution caused by heat radiation and thus obtain the heat radiation received by humans at different evolution times. Then the death probability of humans exposed to heat radiation can be calculated by using the exposure time and the received heat radiation (Q), as follows:

$$Y_f = -14.9 + 2.56\ln\left(6 \times 10^{-3} \times Q^{1.33} \times t_e\right) \tag{4.6}$$

Y_f denotes the probit value with respect to the heat radiation generated by fire. However, the heat radiation received by humans (Q) may vary with the evolution of domino effects since the number of hazardous installations in the "fire" state may change during the evolution. Therefore, the hazardous effects caused by heat radiation at different periods should be superimposed to determine human vulnerability. As a result, at the evolution time T_g, the probit value (Y_{f,T_g}) can be obtained, as follows:

$$Y_{f,t_G} = -14.9 + 2.56\ln\left(6 \times 10^{-3} \times \sum_{g=1}^{g=G} \left(Q_g^{1.33} \times t_g\right)\right) \tag{4.7}$$

The number of people at risk (exposed to accidental scenarios) can also change with the expansion of the accidental area due to domino effects.

4.2.3 Evolution Time

T represents the evolution time of a domino effect event. T is equal to zero when a loss of containment of hazardous materials emerges. The DGMC is divided into G static graphs (G nodes) due to the evolution of domino effects. The parameters of DGMC are updated at each time node T_g due to the change of ignition states, human states, and installation states. In the initial stage, the evolution time mainly depends on the ignition time (*IT*). *IT* is an uncertain variable determined by the number of ignition sources, ignition effectiveness, and vapor cloud dispersion, etc. The *IT* is equal to zero if the released materials are immediately ignited, while it equals the delayed time when it is a delayed ignition. If there is no ignition, the *IT* is infinite. The likelihood of immediate ignition mainly depends on the autoignition of flammable substances and the static discharge caused by the release [47]. If the ignition is delayed, the likelihood of ignition caused by a single ignition source is determined by the ignition effectiveness and the period that the ignition source is covered by the flammable vapor. If there are multiple ignition sources at different positions, the possible ignitions caused by various ignition sources can be considered to be independent events. Thus, the total ignition probability caused by multiple ignition sources can be obtained, as shown in Chap. 3.

4.3 Graph Update Rules and Simulation Algorithm

4.3.1 Graph Update Rules

According to the DGMC model, the graph update rules should be determined to develop the simulation algorithm. The possible state transitions and physical effects in the DGMC model are illustrated in Fig. 4.1.

The dotted lines denote the state transition of nodes (e.g., installations, ignition sources, and humans), while solid lines represent the hazardous effects caused by a node to other nodes. Due to any intentional or unintentional causes, a loss of containment of hazardous materials from an installation may occur, resulting in the state of the installation changes from "operational" to "release". The installation in the "release" state can transfer to a "fire" state if the released materials are ignited immediately. But it may transfer to a "VCE" state if the ignition is delayed. Besides,

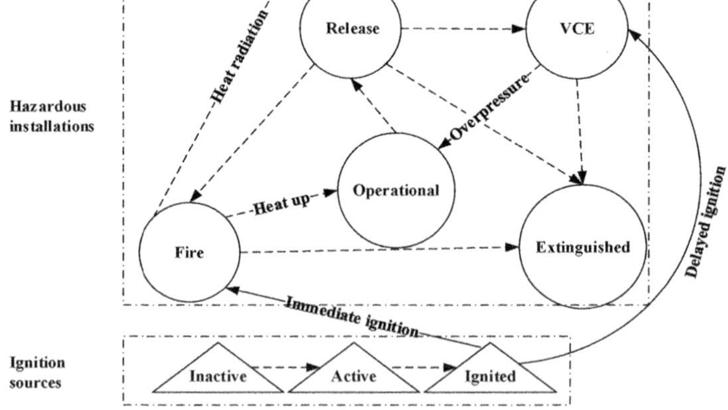

Fig. 4.1 State transition and hazardous effects (Chen et al. [1])

Humans may be injured due to the exposure to toxic gases released from installations, heat radiation generated by installations on fire, and overpressure induced by VCEs caused by a delayed ignition. Moreover, people may die due to prolonged exposure to heat radiation or toxic gases, while they may immediately die due to overpressure because the death likelihood caused by explosions is only determined by the overpressure intensity without the exposure time. In a coupled domino effect, humans may suffer from different accident scenarios at different evolution times and different positions.

To further illustrate the graph update rules, a simple example of a DGMC model with nine static graphs is given in Fig. 4.2. As shown in Fig. 4.2a, the state of the hazardous installation T1 changes from "operational" to "toxic release" when a toxic material releases from an installation. If the released material is ignited immediately, the installation's state will change to "fire" and induce heat radiation on humans and other installations. Otherwise, a flammable vapor cloud may form and disperse along with the vaporization of the release material, resulting in acute toxicity of humans (as shown in Fig. 4.2b). With the dispersion of vapor cloud, the ignition source S1 is covered by the flammable gas, and its state changes from an "inactive" state to an "active" state, as shown in Fig. 4.2c. A VCE occurs when the vapor cloud is ignited by S1, resulting in the damage of hazardous installations and casualties. As shown in Fig. 4.2d, a loss of containment scenario occurs at T2 and T4 while T3 is not damaged by the overpressure. Besides, Humans at H1 and H2 are exposed to the overpressure caused by the VCE. They are injured, but H1 may be more severe than that of H2 due to the synergistic effect caused by the exposures to the toxic gas and the overpressure at H1. Simultaneously, T2 and T4 are on fire due to an immediate ignition, as shown in Fig. 4.2e. In other words, the possible time delay between the explosion (Fig. 4.2d) and fire (Fig. 4.2e) is not considered in this study. In terms of the subsequent scenarios of VCE, they are more likely to be on fire since the explosion can release a lot of heat and energy, increasing the likelihood of immediate ignition at the two damaged tanks. Next, synergistic effects are induced by the two tanks in "fire" states on tanks H1, H2, and T3. As shown in Fig. 4.2f, humans at H1 die at time t_5 due to the accumulation of exposure to heat radiation. Similarly, H2 dies at evolution time t_6, as shown in Fig. 4.2g. At t_7 (Fig. 4.2h), the state of T4 transfers from "fire" to "extinguished" due to the burn-out of flammable substances or firefighting actions. Finally, the fire at T2 is also extinguished, and thus T3 survives since no escalation vectors exist, as shown in Fig. 4.2i.

4.3.2 Simulation Algorithm

The dynamic evolution of a coupled domino effect is represented by a dynamic graph (see Fig. 4.2). Due to the evolution of uncertainties such as the ignition time and the accident scenarios in coupled domino effects, numerous dynamic graphs should list all the possible evolution paths. Besides, more dynamic graphs are needed for real chemical industrial areas with multiple ignition sources and a vast variety of

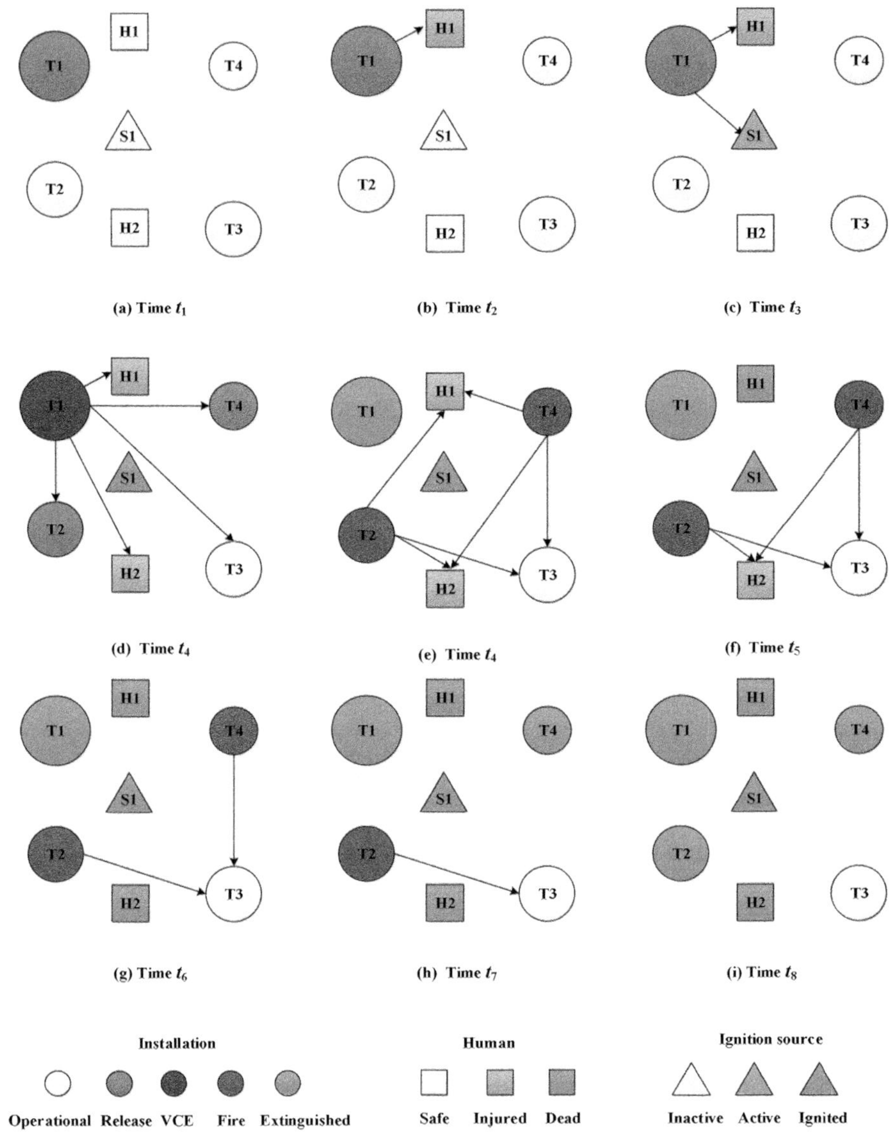

Fig. 4.2 A DGMC with nine static graphs (Chen et al. [1])

hazardous installations. To be able to cope with this situation, an algorithm based on Monte Carlo simulation is developed to generate dynamic graphs, tackling the time-dependencies and uncertainties in coupled domino effects. Figure 4.3 shows the developed algorithm based on the Monte Carlo simulation for solving the DGMC model.

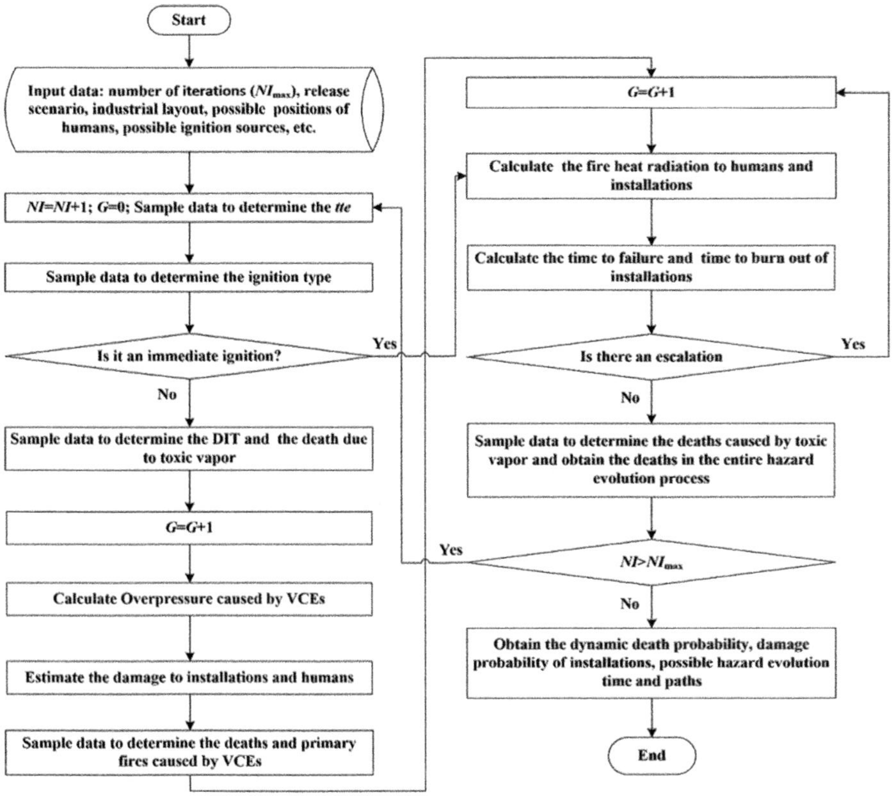

Fig. 4.3 Simulation algorithm based on Monte Carlo method (Chen et al. [1])

As shown in Fig. 4.3, the first step of the algorithm is to input all the needed basic data related to the chemical plant, such as the number of iterations (NI_{max}), the industrial layout, release scenarios, possible ignition sources, etc. Step 2 determines the ignition type using random sampling. If it is a delayed ignition, the ignition time (*IT*) and the ignition source are determined by random sampling; otherwise, the ignition time equals zero. Then the death caused by the exposure to toxic gases can also be determined using Eq. (4.2). At the *IT*, the dynamic graph should be updated, and the edges in the graph represent overpressure induced by the VCE. The death probability caused by overpressure for humans and hazardous installations can be calculated according to Eq. (4.3). The fatalities and damaged installations are determined by using random sampling. The graph will be updated again when a fire occurs, and the edges in the graph represent heat radiations from the installation on fire to other installations. Next, the time to failure of hazardous installations can be determined using the method illustrated in Chap. 2. If there is an immediate ignition, the calculation procedures for explosion and toxic vapor can be overlooked, and only the calculation related to fire escalation is conducted. During the fire escalation period, the graph is updated when a new fire occurs or an existing fire is extinguished when the evolution is over.

The above calculation steps will be repeated NI_{max} times. Finally, the death probability of humans and failure probability of installations during the dynamic evolution of coupled domino effects are obtained. Besides, we can also obtain the possible evolution paths, evolution time nodes, and ignition times by using the algorithm.

4.4 Case Study

To illustrate the approach, a chemical storage facility with 37 chemical storage tanks (T1–T37), 5 possible human positions (H1–H5), and two possible ignition sources (S1–S2) is used as a case study. Table 4.5 lists the main features of storage tanks. The layout of the storage plant is shown in Fig. 4.4.

An overfilling of acrylonitrile at T1 with a filling rate of 100 kg/s is considered the primary scenario. The release of acrylonitrile can result in acute intoxication and

Table 4.5 Features of storage tanks (Chen et al. [1])

Tank	Dimension × Height (m)	Chemical substance	Nominal volume (m^3)	Chemical content (m^3)
T1–T6	21.0 × 16.6	Acrylonitrile	5000	4000
T7–T9, T12–T15	17.0 × 15.4	Gasoline	3000	2400
T10, T11	7.0 × 13.6	Gasoline	500	400
T16–T27	14.5 × 12.7	Gasoline	2000	1600

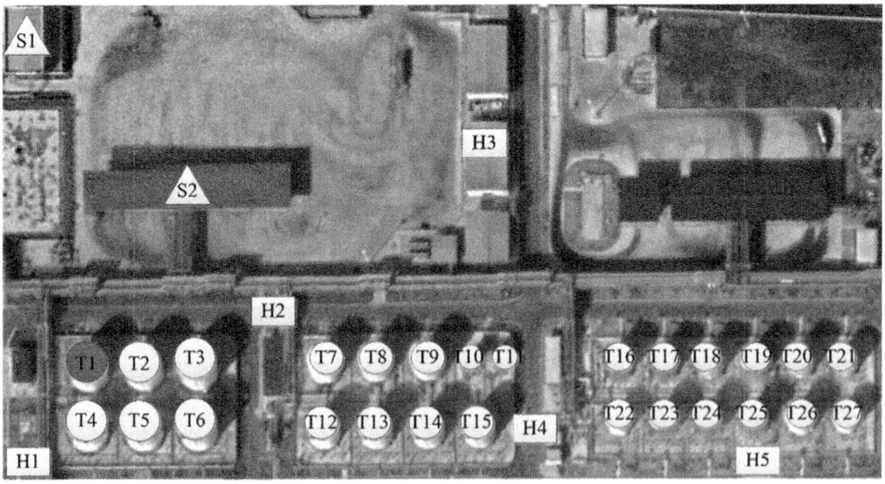

Fig. 4.4 Chemical storage facility considered in the case study (Chen et al. [1])

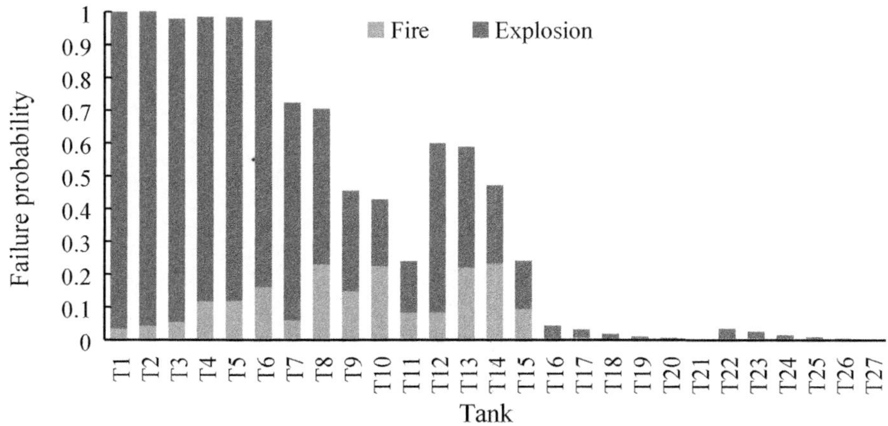

Fig. 4.5 Failure probabilities of tanks caused by fire and explosion (Chen et al. [1])

possible subsequent explosions and fires may lead to overpressure and heat radiation. The ambient temperature is 0 °C and the wind speed is equal to zero. According to the vapor cloud dispersion model, the ignition source S1 is in the "active" state after 5.1 min and S2 is in the "active" state after 2.8 min. The autoignition probability P_{ia} is zero and the ignition probability due to static discharge P_{is} is 0.02, the autoignition temperature is 481 °C [48]. The possible heat radiations induced by tanks and the burning rates of fires are calculated through the ALOHA software. The number of iterations ($NI_{\max}$) is set as 10^5 and the computation time is 3.9 min.

Figure 4.5 shows the failure probabilities caused by possible fires and explosions due to the overfilling of acrylonitrile at T1. The failure probabilities of T1-T6 are near 1 since they are close to the release tank. The failure probabilities of tanks obviously decrease with increasing the distance between the release sources and the tanks (e.g., T7–T11, T12–T15). Based on the probabilities caused by fire and explosion of each tank, the tanks close to T1 (i.e., T1–T8 and T12) are more likely to be directly damaged by VCEs. These damaged tanks may catch fires and escalate spatially as well as temporally, resulting in the damage of other tanks such as T10 and T11.

Due to the spatial–temporal evolution of accident scenarios, people may suffer from different hazardous accident scenarios. The death probabilities caused by different scenarios at H1–H5 are shown in Fig. 4.6. The total death probabilities at H1 and H2 are around 0.99, indicating that people at H1 and H2 are very likely to die due to exposure to toxic gas or fire. The death probability caused by fire and toxic gas at H1 and H2 are approximately identical, which indicates that people at H1 and H2 may simultaneously or sequentially receive the impact from multiple accident scenarios. H3 has a lower death probability caused by toxic gas because it is located far away from T1 and it is also not close to any of the storage tanks. Contrarily, fire can escalate from tanks nearby T1 to the tanks close to H5 although there is a long distance between T1 and H5. As a result, H5 may die due to fire and

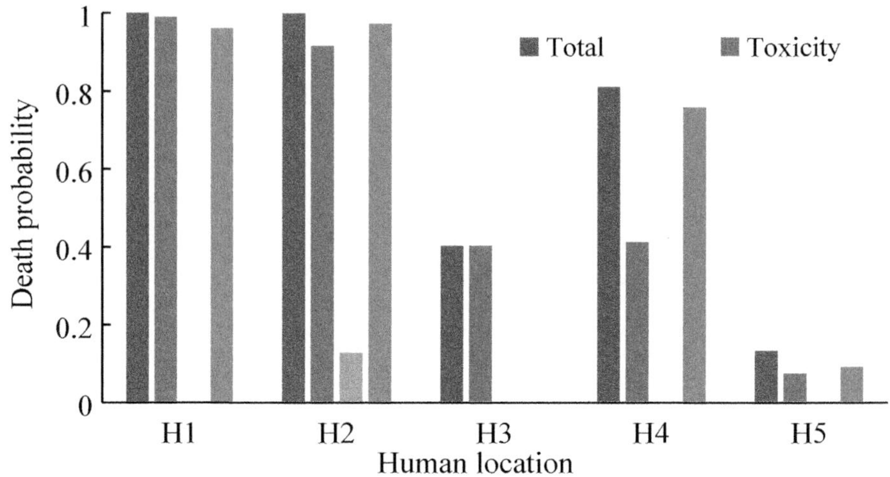

Fig. 4.6 Death probabilities caused by different accident scenarios at H1–H5 (Chen et al. [1])

toxic gas. According to these results, different personal protective equipment (PPE) can be assigned to people in different locations or at different times. For instance, people at H3 only need to take a gas mask while protective clothing against potential heat is also needed for people at H1, H2, H4, and H5.

4.5 Conclusions

This chapter develops a dynamic graph Monte Carlo (DGMC) model to model the evolution of coupled hazardous scenarios and assess the vulnerability of humans and facilities. The DGMC model can effectively model simultaneous or sequential multiple hazardous scenarios caused by coupled domino effects. Suppose only one type of scenario is considered in coupled domino effects. In that case, the vulnerability of installations and humans may be primarily underestimated, resulting in the unreasonable allocation of personal protection equipment (PPE) and safety barriers. Since humans in different positions may suffer from different hazardous scenarios, different protection strategies should be implemented for people in different positions; According to the application of the developed model [1], VCEs may result in the damage of multiple hazardous installations and the safety distances based on fire hazards may not be sufficient for VCEs; Furthermore, humans close to the release source are prone to multiple hazardous scenarios while acute toxicity and VCEs are the main causes for the deaths outside the release areas.

References

1. Chen C, Reniers G, Khakzad N (2021) A dynamic multi-agent approach for modeling the evolution of multi-hazard accident scenarios in chemical plants. Reliab Eng Syst Saf 207
2. Chen C (2021) A dynamic and integrated approach for modeling and managing domino-effects. Delft University of Technology
3. Chen C, Reniers G, Khakzad N (2019) Integrating safety and security resources to protect chemical industrial parks from man-made domino effects: a dynamic graph approach. Reliab Eng Syst Saf 191
4. Wang B, Li D, Wu C (2020) Characteristics of hazardous chemical accidents during hot season in China from 1989 to 2019: a statistical investigation. Saf Sci 129
5. Yang Y, Chen G, Reniers G (2019) Vulnerability assessment of atmospheric storage tanks to floods based on logistic regression. Reliab Eng Syst Saf 106721
6. Wang B, Wu C, Reniers G, Huang L, Kang L, Zhang L (2018) The future of hazardous chemical safety in China: opportunities, problems, challenges and tasks. Sci Total Environ 643:1–11
7. Vilchez JA, Sevilla S, Montiel H, Casal J (1995) Historical analysis of accidents in chemical plants and in the transportation of hazardous materials. J Loss Prev Process Ind 8(2):87–96
8. Mishra KB, Wehrstedt K-D, Krebs H (2013) Lessons learned from recent fuel storage fires. Fuel Process Technol 107:166–172
9. Mishra KB, Wehrstedt K-D, Krebs H (2014) Amuay refinery disaster: the aftermaths and challenges ahead. Fuel Process Technol 119:198–203
10. Maremonti M, Russo G, Salzano E, Tufano V (1999) Post-accident analysis of vapour cloud explosions in fuel storage areas. Process Saf Environ Prot 77(6):360–365
11. Taveau J (2012) The Buncefield explosion: were the resulting overpressures really unforeseeable? Process Saf Prog 31(1):55–71
12. Sharma RK, Gurjar BR, Wate SR, Ghuge SP, Agrawal R (2013) Assessment of an accidental vapour cloud explosion: lessons from the Indian Oil Corporation Ltd. accident at Jaipur, India. J Loss Prev Process Ind 26(1):82–90
13. Dasgotra A, Varun Teja GVV, Sharma A, Mishra KB (2018) CFD modeling of large-scale flammable cloud dispersion using FLACS. J Loss Prev Process Ind 56:531–536
14. Mishra KB (2018) The influence of volume blockage ratio on IOCL Jaipur explosion. J Loss Prev Process Ind 54:196–205
15. Gant SE, Atkinson GT (2011) Dispersion of the vapour cloud in the Buncefield Incident. Process Saf Environ Prot 89(6):391–403
16. Salzano E, Cozzani V (2003) The use of probit functions in the quantitative risk assessment of domino accidents caused by overpressure. Saf Reliab 1 and 2. A Balkema Publishers, Leiden
17. Cozzani V, Salzano E (2004) The quantitative assessment of domino effects caused by overpressure: Part I. Probit models. J Hazard Mater 107(3):67–80
18. Cozzani V, Salzano E (2004) The quantitative assessment of domino effect caused by overpressure: Part II. Case studies. J Hazard Mater 107(3):81–94
19. Zhang M, Jiang J (2008) An improved probit method for assessment of domino effect to chemical process equipment caused by overpressure. J Hazard Mater 158(2–3):280–286
20. Mukhim ED, Abbasi T, Tauseef SM, Abbasi SA (2017) Domino effect in chemical process industries triggered by overpressure—formulation of equipment-specific probits. Process Saf Environ Prot 106:263–273
21. Zhou J, Reniers G (2017) Petri-net based cascading effect analysis of vapor cloud explosions. J Loss Prev Process Ind 48:118–125
22. Yang Y, Chen G, Chen P (2018) The probability prediction method of domino effect triggered by lightning in chemical tank farm. Process Saf Environ Prot 116:106–114
23. Khakzad N, Landucci G, Reniers G (2017) Application of graph theory to cost-effective fire protection of chemical plants during domino effects. Risk Anal 37(9):1652–1667
24. Khakzad N, Reniers G, Abbassi R, Khan F (2016) Vulnerability analysis of process plants subject to domino effects. Reliab Eng Syst Saf 154:127–136

25. Landucci G, Gubinelli G, Antonioni G, Cozzani V (2009) The assessment of the damage probability of storage tanks in domino events triggered by fire. Accid Anal Prev 41(6):1206–1215
26. Khakzad N (2015) Application of dynamic Bayesian network to risk analysis of domino effects in chemical infrastructures. Reliab Eng Syst Saf 138:263–272
27. Chen C, Reniers G, Zhang L (2018) An innovative methodology for quickly modeling the spatial-temporal evolution of domino accidents triggered by fire. J Loss Prev Process Ind 54:312–324
28. Kamil MZ, Taleb-Berrouane M, Khan F, Ahmed S (2019) Dynamic domino effect risk assessment using Petri-nets. Process Saf Environ Prot 124:308–316
29. Zeng T, Chen G, Yang Y, Chen P, Reniers G (2019) Developing an advanced dynamic risk analysis method for fire-related domino effects. Process Saf Environ Prot
30. Zhang L, Landucci G, Reniers G, Khakzad N, Zhou J (2018) DAMS: a model to assess domino effects by using agent-based modeling and simulation. Risk Anal 38(8):1585–1600
31. Abdolhamidzadeh B, Abbasi T, Rashtchian D, Abbasi SA (2010) A new method for assessing domino effect in chemical process industry. J Hazard Mater 182(1–3):416–426
32. Rad A, Abdolhamidzadeh B, Abbasi T, Rashtchian D (2014) FREEDOM II: an improved methodology to assess domino effect frequency using simulation techniques. Process Saf Environ Prot 92(6):714–722
33. Bonabeau E (2002) Agent-based modeling: methods and techniques for simulating human systems. Proc Natl Acad Sci 99(suppl 3):7280–7287
34. Rai S, Hu X (2018) Hybrid agent-based and graph-based modeling for building occupancy simulation. In: Proceedings of the 4th ACM international conference of computing for engineering and sciences, pp 1–12
35. Stroeve SH, Blom HA, Bakker GB (2013) Contrasting safety assessments of a runway incursion scenario: event sequence analysis versus multi-agent dynamic risk modelling. Reliab Eng Syst Saf 109:133–149
36. Harary F (1969) Graph theory. Addison-Wesley, Reading, MA
37. Jafari S, Ajorlou A, Aghdam AG (2011) Leader localization in multi-agent systems subject to failure: a graph-theoretic approach. Automatica 47(8):1744–1750
38. Ding L, Ji J, Khan F (2020) Combining uncertainty reasoning and deterministic modeling for risk analysis of fire-induced domino effects. Saf Sci 129
39. Jiang D, Wu B, Cheng Z, Xue J, van Gelder PHAJM (2020) Towards a probabilistic model for estimation of grounding accidents in fluctuating backwater zone of the Three Gorges Reservoir. Reliab Eng Syst Saf
40. Khakzad N, Reniers G (2015) Using graph theory to analyze the vulnerability of process plants in the context of cascading effects. Reliab Eng Syst Saf 143:63–73
41. Rubinstein RY, Kroese DP (2016) Simulation and the Monte Carlo method, vol 10. Wiley
42. Joy DC (1995) Monte Carlo modeling for electron microscopy and microanalysis, vol 9. Oxford University Press
43. Kuczera G, Parent E (1998) Monte Carlo assessment of parameter uncertainty in conceptual catchment models: the Metropolis algorithm. J Hydrol 211(1–4):69–85
44. Van Den Bosh C, Merx W, Jansen C, De Weger D, Reuzel P, Leeuwen D, Blom-Bruggerman J (1989) Methods for the calculation of possible damage (Green Book). Committee for the Prevention of Disasters, The Hague (NL)
45. Uijt de Haag, Ale (1999) Guidelines for quantitative risk assessment. Committee for the Prevention of Disasters, The Hague (NL)
46. Atkinson G, Coldrick S (2012) Vapour cloud formation: experiments and modelling. Health and Safety Laboratory, Debyshire
47. Chen C, Khakzad N, Reniers G (2020) Dynamic vulnerability assessment of process plants with respect to vapor cloud explosions. Reliab Eng Syst Saf 200
48. Brazdil JF (2000) Acrylonitrile. Ullmann's encyclopedia of industrial chemistry

Chapter 5
Integrated Safety and Security Management to Tackle Domino Effects

5.1 Introduction

A series of risk assessment methods for domino effects in the chemical industry are introduced in Chaps. 2, 3 and 4. These methods can model both unintentional domino effects and intentional domino effects.[1] To prevent and mitigate domino effects, domino effect management methods should be developed based on the results delivered by domino effect risk assessments. Previous work on domino effect management mainly focused on safety barriers management and land-use planning to prevent domino effects. Landucci et al. [3] explored the role of safety barriers in managing domino effects. Janssens et al. [4] developed an optimization method to allocate protective safety barriers and mitigate domino effects in chemical plants. Ghasemi and Nourai [5] optimized the spacing of storage tanks in land use planning for minimizing domino effect risks. Khakzad and Reniers [6] established a cost-effectiveness approach for the allocation of safety measures in chemical plants for land-use planning. These attempts on domino effect management aim to prevent and mitigate domino effects induced by unintentional events. However, the consequences of intentional domino effects may be more severe due to for instance simultaneous attacks on multiple chemical installations.

Besides safety barriers, security measures may also be used to prevent possible multiple primary scenarios. Process security in the chemical industry is a relatively new domain compared to process safety. The U.S. Department of Homeland Security identified the chemical sector as one of 16 critical infrastructures vulnerable to intentional attacks [7]. An increasing public concern raised the attention on process security after the terrorist attack in New York City on September 11, 2001 [8–12]. In 2003, the Center for Chemical Process Safety (CCPS) issued guidelines for analyzing and managing the security vulnerabilities of chemical industrial areas. The guidelines provide useful tools to industrial companies that deal with hazardous materials for

[1] This chapter is mainly based on two publications: Chen et al. [1] and Chen [2].

C. Chen et al., *Integrating Safety and Security Management to Protect Chemical Industrial Areas from Domino Effects*, Springer Series in Reliability Engineering,
https://doi.org/10.1007/978-3-030-88911-1_5

assessing and managing their risks caused by terrorists. A comprehensive approach was developed to integrate process security management and process safety management strategies. However, possible domino effects triggered by (unintentional or) intentional events were ignored in this document [13].

In 2004, the American Petroleum Institute (API) published a recommendation on security risk assessment for the petroleum and petrochemical industries [14]. This document provides a systematic security risk assessment (SRA) method based on threat, vulnerability, and consequence analysis. In 2013, the SRA method was revised by expanding functional utility [15]. According to the SRA method, the security risk is a function of threat, vulnerability, and consequence. Threat analysis is a considerable challenge since it requires a multitude of data and knowledge and modeling the motivations, intents, characteristics, capabilities, and tactics of adversaries [16, 17]. Vulnerability analysis requires a detailed understanding of the design and operation of installations and the threat information [18–20]. Security measures can improve the capability of installations against attacks but may change the attackers' strategies because of the intelligent character of the adversaries [21–23]. Consequently, Game theory was recognized as a promising tool for analyzing adversaries' strategies and optimizing the defenders' response via optimal allocation of security resources [24–27].

In terms of intentional domino effects, Reniers and Audenaert [28] proposed using safety barriers to mitigate the potential consequences of intentional attacks. However, the escalation caused by intentional events may be very difficult to prevent due to possible acceleration induced by synergistic effects. Consequently, integrating safety and security management may be a viable approach to deal with intentional and unintentional domino effects. However, there is a research gap to integrate safety and security management to prevent and mitigate domino effects in chemical industrial areas. This chapter, therefore, provides a framework to this end.

5.2 Safety and Security Management Principles

5.2.1 Inherent Safety and Security

In light of the severe consequences of the Flixborough disaster in 1974 and the Bhopal disaster in 1984, Kletz [29] proposed the concept of inherent safety or inherently Safer design (ISD) to make chemical plants safer. Unlike traditional plant designs that try to reduce the risk by adding protective equipment and following safety procedures, inherent safety aims to remove or reduce the hazards from the start, hence, by design. In terms of security, inherent safety also leads to the removal or reduction of threats. If hazards or threats are removed, no undesired events are possible and thus no subsequent domino effects are possible. Inherent security should thereby also reduce the attractiveness of assets belonging to chemical plants and therefore also remove or reduce threats. Kletz and Amyotte [30] proposed 10 inherent safety principles:

- **Intensification or minimization**: Reduction of the number and amount of hazardous materials in a chemical plant.
- **Substitution**: Substitution of a hazardous material by a safer one; For instance, a chemical plant can use a non-flammable material to replace a flammable material.
- **Attenuation**: attenuation of hazardous effects by using a hazardous material under safer conditions such as low temperature and low pressure.
- **Limitation of effects**: limitation of hazardous effects by changing designs (e.g., unit segregation design), operations, or reaction conditions.
- **Simplicity**: Simplicity of equipment, process, operations for reducing opportunities for errors and equipment failure.
- **Avoiding domino effects**: avoiding the escalation of major accidents by increasing layout spacing, providing firebreaks between sections, etc.
- **Making incorrect assembly impossible**: chemical plants should be designed to avoid incorrect assembly. For instance, a compressor valve should be designed to avoid incorrect assembly of inlet and exit valves.
- **Making status clear**: avoiding complicated equipment and overloading information and marking important information.
- **Tolerance of misuse**: tolerance of poor installation or operation without failure. For example, expansion loops are more tolerant of poor installation than bellows in the pipework.
- **Ease of control**: control by using physical principles rather than adding control equipment and avoidance of hands-on control [31].

The above principles are guidelines for an inherent safety (and security) design from a safety perspective. Some unique principles for inherent security are proposed in this book, as follows:

- **Reduction of attractiveness**: reduction of the assets that are attractive for adversaries;
- **Hidden attractive assets**: attractive assets should be located in hidden places or covered by other materials to make them difficult to be found.
- **Making the access of attractive assets difficult**: Attractive assets should be designed to be difficult to assess. For example, the control center is located far from the entrances.
- **Least privilege**: Each attractive asset or part of the digital system is designed with the privileges needed for its function to avoid multiple losses of assets or the failure of the whole digital system.

With the application of inherent safety and security design principles, it is expected to achieve fewer hazards, threats, fewer additional safety and security measures, fewer people to be exposed to hazardous effects, and fewer attractive assets to be exposed to threats.

5.2.2 Layers/Rings of Protection

CCPS [32] defined an independent protection layer (IPL) as a device, system, or action that can prevent a scenario from proceeding to undesired consequences, independent of the initiating event and other protection layers for the scenario. The concept of rings of protection (RoP) in the security domain defines a protection system as a series of independent protection rings [13]. It is similar to the layers of protection (LOP) used in chemical process plants for safety barrier management [33]. The concept of "Layers of Protection" is usually used for safety barrier management in chemical process plants [13] and the "Rings of Protection" for security are derived from this concept. A layer of protection analysis (LOPA) is a simple risk assessment method used for judging whether there are sufficient protection layers to control the risk of an undesired event.

Protection layers are independent of each other and not influenced by the likelihood or consequences of the initiating event [32]. The performance of a protection layer depends on the availability and effectiveness of safety barriers associated with the layer. Availability denotes the probability of failure on demand (PFD) of the safety barriers, while effectiveness is the probability that the safety barrier can prevent the scenario if it is successfully activated. Since no layer is perfectly effective, a scenario may require more than one protection layer depending on the process complexity and potential severity of a consequence [32]. For example, a pressure relief valve, a fire sprinkler system, a fireproofing coating, and a firefighting team may be simultaneously used for preventing fire escalation and failure of hazardous material vessels, as shown in Fig. 5.1. Each safety barrier in Fig. 5.1 is an independent protection layer. The escalation risk decrease with the increase of protection layers. If the protection layers are sufficient to prevent the escalation, no additional protection layers are needed. Otherwise, more protection layers should be implemented.

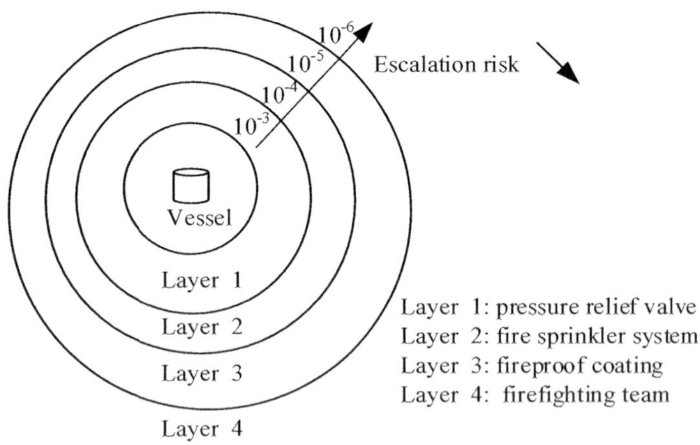

Fig. 5.1 Layers of protection: an illustrative example

If the risk tolerance criterion for domino effects is determined in a chemical plant, appropriate protection layers should be used to decrease the escalation risk below the tolerance criterion. Besides, the escalation risk can also be reduced by improving the availability and effectiveness of safety barriers in each layer.

In terms of security, the concept of "Rings of Protection" is also known as "Layered Defenses" and is based on the "Defense in Depth" principle [34]. Each defense layer consists of any device, system, or action that is capable of preventing the success of an attack. Appropriate defense layers need to be implemented in a chemical plant to protect critical assets. In terms of domino effects, critical installations may be defined as any hazardous facility with a high probability of initiating or propagating domino effects. If the current security risk is higher than the tolerance criteria, additional protection layers may be needed as attack prevention or mitigation measures. In chemical plants, security barriers such as a locked door and a fence can be considered to be preventive rings, while safety barriers such as a fire sprinkler that can prevent possible escalation caused by attacks may be regarded as mitigation rings.

According to the theory of protection rings, the most critical or vulnerable assets (e.g., the control room and hazardous material storage tanks) should be put into the center of concentric levels. In that case, an attack has to penetrate many defense rings, such as a fence, a locked exterior door, an alert receptionist, an elevator with key-controlled floor buttons, and a locked door [13]. It should be remarked that the defense against external attacks is less difficult than that against attacks within the company because internal attackers can easily bypass or penetrate part of the rings [35]. Unlike protection layers of safety barriers, the effectiveness of a protection ring constituting security measures is influenced by adversaries. If adversaries have the protection information of a chemical plant, they may use available means to lower the availability of protection rings. These means may for instance include using explosions to induce the failure of protection devices, releasing toxic gases to incapacitate all inhabitants of the control room, or bypass defense layers. As a result, defense rings should be robust, and the details of protection information may not be open to the public.

5.3 Integrated Safety and Security Management

5.3.1 Motivations for Integrating Safety and Security

According to the analysis in Sect. 5.2, both safety management and security management can be achieved by the inherent safety and security principles and the layers/rings of protection. In other words, both safety and security can be integrated into the design stage and operation stage in light of the intentional and unintentional threats associated with chemical plants. Safety barriers for preventing domino effects in process industries are generally divided into three categories: (i) active protection systems, (ii) passive protection systems, and (iii) emergency measures

[33, 36]. Previous researches on the management of domino effects mainly focus on accidental domino effects. Landucci et al. [37] developed a fault tree methodology to quantify the performance of safety barriers in fire-induced domino effects. Janssens et al. [4] developed an optimization model to allocate safety barriers for the sake of maximizing the *ttf* of chemical installations. Khakzad et al. [38] proposed a DBN approach for the performance assessment of fire protection systems during domino effects, taking into account the dynamic failure process of fireproofing coatings. Khakzad et al. [39] also developed an approach based on a limited memory influence diagram (LIMID) to multi-attribute decision analysis of safety measures. Advanced tools such as Petri-net and event sequence diagrams were applied to assess emergency response actions during fire-induced domino effects [40, 41].

A few attempts to manage intentional domino effects are based on securing critical installations or reducing potential consequences using safety barriers, ignoring the integration of safety barriers and security measures. Figure 5.2 shows the dependencies in the decision-making on safety and security resources, assuming the possibility of domino effects occurring after the allocation.

Safety barriers can prevent or mitigate accidental domino effects and also may have an impact on intentional domino effect scenarios. For instance, safety barriers may reduce the potential consequences of intentional events and decrease the attractiveness of chemical industrial areas. In terms of cross-plant areas, safety and security resources allocated in one chemical plant may lead to benefit of plants nearby due to the mitigation of possible external domino effects while it may also relatively increase the security risk of nearby plants because of the change of attractiveness. Therefore, safety and security resources should be integrated to prevent or mitigate

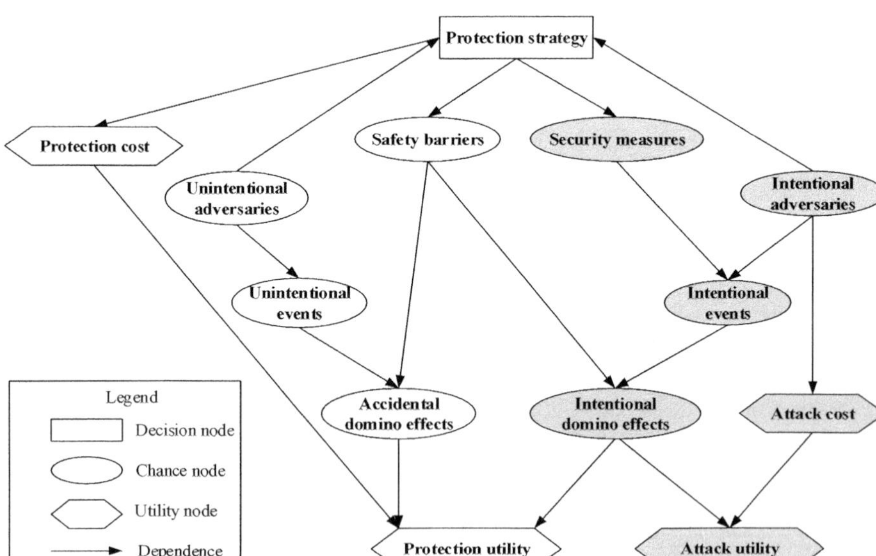

Fig. 5.2 The diagram for the allocation of safety and security resources

all possible domino effects caused by safety or security accident scenarios. In this book, "integrated safety and security measures" are called protection measures.

5.3.2 *Classification of Protection Measures*

In light of possible intentional and accidental domino effects, both safety and security measures may be used in domino effect management. A new classification of protection measures (safety or security measures) for preventing and mitigating domino effects in chemical industrial areas is proposed, as shown in Fig. 5.3.

According to their functions, protection measures are divided into three categories: detection measures, delay barriers, and emergency response actions. The three types of protection measures will be explained and elaborated below.

(1) **Detection measures**

Detection measures are used to detect intentional and unintentional abnormal events such as accidental releases and adversary actions and take the necessary actions to deal with them. The detection function consists of three sub-functions: (i) sensing, seeing, and discovering the problems (sensor), (ii) evaluating, assessing, and thinking about a solution for the problem (logic solution), and (iii) doing, acting, and carrying acts to solve the problem (actuator). A detection for an abnormal event is successful only when the functions are all there correctly executed. As a result, the detection performance depends on the probability that detection sensors or persons successfully

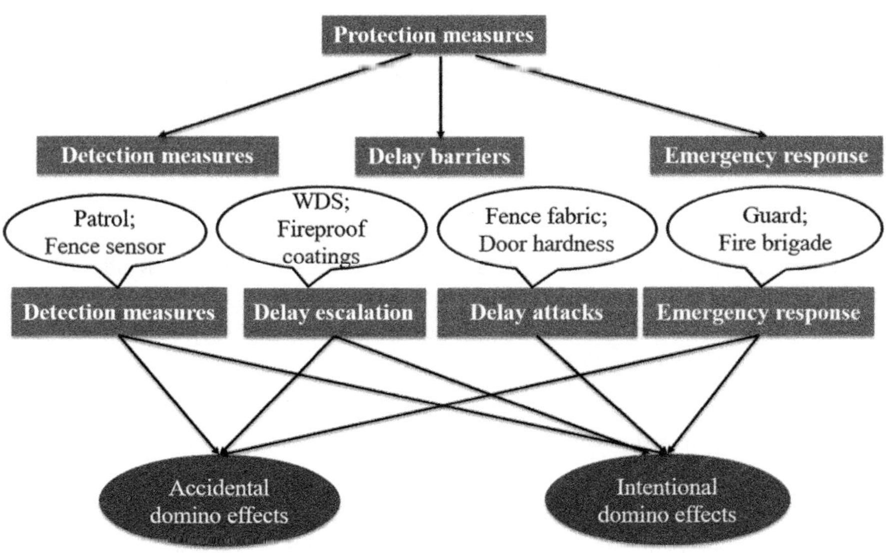

Fig. 5.3 Classification of add-on protection measures related to domino effect management

discover abnormal events, the probability that the alarm related to the events is successfully communicated, the probability that the alarm is successfully assessed and acted upon, and the time needed for the entire detection process (detection time) is adequate. An additional detection performance indicator is the nuisance alarm rate. A nuisance alarm is any alarm that is not caused by abnormal events. In an ideal detection system, the detection probability would be 1, and the nuisance alarm rate would be zero. However, in a chemical plant, all sensors interact with their environment, and they may be disturbed by other disturbances in their detection zones, such as vegetation, wildlife, and weather conditions. In a chemical plant, a typical detection system may consist of exterior and interior intrusion sensors, video alarm assessment, entry control, and alarm communication systems. Video alarm assessment is always conducted by closed-circuit television (CCTV) camera coverage of each sensor sector. An entry control system allows authorized personnel and material to get into and out of facilities while detecting and possibly prevent the access of unauthorized movements. Alarm communication aims to transport an alarm and information to a center and possibly present and assess the information [34, 42].

(2) **Delay barriers**

Delay barriers may be divided into two categories: delay attack measures and delay escalation measures. Delay attack measures are barriers that can increase the time in which an adversary needs to carry out an action, and such measures can thus delay the implementation of an attack. An adequate protection system first requires that the detection system successfully discovers the abnormal event. In that case, response force (e.g., guards, police) can prevent the attack when the alarm is correctly assessed and delivered to the response force. However, the start of an effective response force needs time. If the time is larger than the time required for completing the adversary attack, the response force would be ineffective, and the attack can not be prevented. As a result, after an adversary action is detected, delay measures are employed to delay the implementation of the attack until an adequate response force is available. Therefore, the response force can successfully interrupt the adversary attack before the attack goal is achieved only when the adversary is detected. The response force is available (active) before the attack is implemented [1, 42]. Delay escalation barriers are barriers to delaying the escalation of major accidents, such as fireproof coatings and water delivery systems. For example, fireproof coatings can block heat radiation transfer from the installation on fire to nearby installations exposed to the fire, increasing the time to failure (ttf) of the exposed installations [43, 44].

(3) **Emergency response**

Response force refers to any response personnel and measures involved in response to intentional attacks or hazardous scenarios in a chemical plant. As a result, emergency response in a chemical plant is essential to protect installations, the public, workers, and the environment. The response force may be on-site and offsite, including security guards, police, medical emergency teams, the fire brigade, etc. Guards and police may be regarded as a preventive response that may prevent the completion of an attack if the attack is successfully detected. And if the delay measures provide enough time for the start of the response force. Medical emergency teams and fire brigades are used to mitigate the consequences of attacks. Protection of different targets may require response plans and the performance of response force depends on the threat types. For example, it is almost impossible to use security guards to prevent drone attacks. This is one reason why the drone attack on the Abqaiq oil plant in 2019 led to a 50% reduction in Abqaiq's oil production and a nearly 15% increase in the crude oil price [45]. Therefore, inherent safety and security, besides the add-on protection measures, is very important, and different chemical plants may also need additional protection strategies, and an effective protection strategy requires a reasonable arrangement of detection, delay, and response measures [1, 34, 42].

5.4 An Integrated Approach to Manage Domino Effects

This section develops an integrated management framework for preventing and mitigating intentional and unintentional domino effects. The procedures of the developed approach are shown in Fig. 5.4.

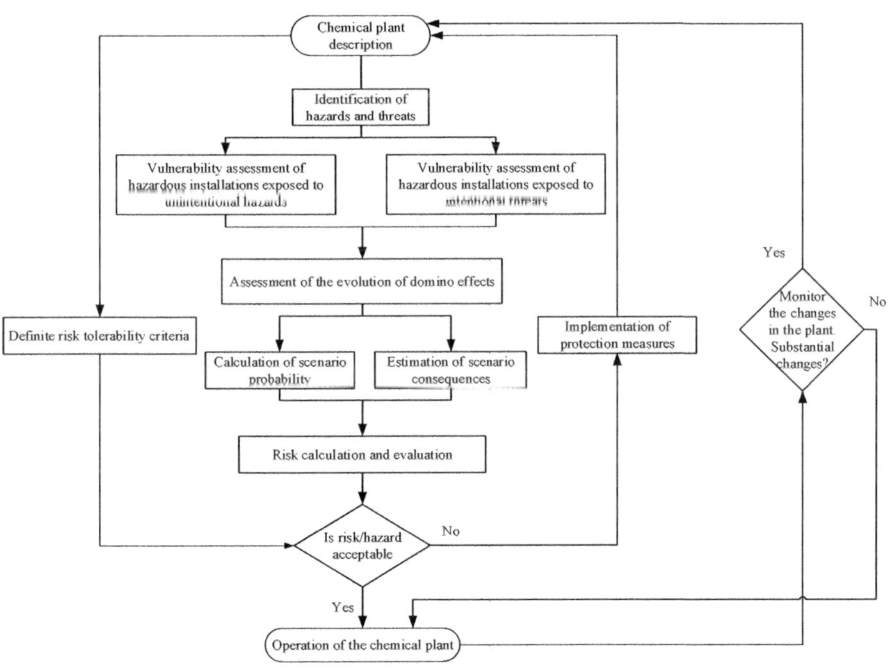

Fig. 5.4 Integrated management framework for domino effects

First, we need to collect all the data and information of a chemical plant related to risk assessment and management. Based on this information and data, possible intentional threats and hazards should be identified for the basis of vulnerability analysis. In this framework, vulnerability assessment should consider (i) the vulnerability of installations directly against threats and hazards and (ii) the vulnerability of installations subject to possible domino effects caused by intentional and unintentional events. Next, a consequence analysis should be conducted to obtain the possible consequences of domino effects. Furtherly, the domino effect risk is calculated and evaluated to judge whether risk reduction is needed by comparing it against the risk tolerability criteria. If the risk is higher than the criteria, risk reduction strategies (add-on safety and security measures) should be formulated.

5.4.1 Chemical Plant Description

In the first step, necessary data and information associated with the scope of the study need to be collected. The foremost necessary information and data are as follows:

- Plant information: The location of the chemical plant, the social, political, and economic environment around the chemical plant, nearby chemical plants or other critical infrastructures, products, etc.
- Layout information: Asset positions, distances between different installations, physical links between various installations, etc.
- Asset information: numbers, types, shapes, sizes, value, functions, materials, etc.
- Hazardous material information: types, quality, locations, hazardous characteristics, states (e.g., phase state, pressure, and temperature).
- Protection measures information: detection measures, delay measures, internal emergency response, and external emergency response.
- Meteorological data: temperature, wind direction, speed, humidity, etc.

5.4.2 Threat and Hazard Analysis

In terms of domino effects, this step needs to answer the question: what can induce primary scenarios? Since domino effects can be caused by intentional attacks and unintentional events, the step of threat and hazard analysis should identify both hazards and intentional threats (that is, the intentional issues of hazards), providing information and data for carrying out a vulnerability analysis.

Hazard identification in a chemical plant is to identify unintentional possible unwanted events leading to losses for people, property losses, and environmental damages, obtaining the weaknesses in the design and operation that could lead to hazardous material releases, fires, or explosions [46]. These hazards are always caused by the presence of hazardous materials or the processing conditions and hence, they belong to the field of "process safety". The characteristics of hazards can be flammability, combustibility, toxicity, corrosivity, radioactivity, etc. The processing conditions refer to hazardous physical conditions such as high temperature, high pressure, vibration, and liquid hammering and the strike [46]. These unintentional threats can be analyzed by widely used hazard identification methods such as the Hazard and Operability Study (HAZOP) method and the checklist method.

An intentional threat may be regarded as an indication, a circumstance, or an event that can lead to deliberate damage to humans, property and the environment, etc. [15]. Intentional threat analysis should identify possible threats and the attractiveness of installations to these threats. Intentional threat identification needs to collect information and data on potential threats, such as threat types, motivations, and weapons. The types of adversaries in chemical plants include terrorists, criminals, violent activities, deranged individuals, and disgruntled employees [13]. The motivations of threats may be the willingness to die, maximizing damage and casualties, inflicting psychological terror on the employees and public, and demonstrating the inability of the host company or country for instance [13]. Adversaries may take simple physical actions (e.g., opening valves, cutting electricity cables, and distorting production parameters), use simple arms (e.g., handguns, knives, and explosives), and exploit advanced weapons (missiles, drones, and nuclear weapons) to increase the likelihood of a successful attack and expand the consequences [13]. Adversaries may be internal adversaries, external adversaries, or internal adversaries working in cooperation with external adversaries, and they may be individuals or from groups, organizations, or governments. Threat analysis should consider as many adversaries as possible, such as intelligence services of host nations or third-party nations, political and terrorist groups, criminals, rogue employees, cybercriminals, and private interests [15]. Based on the information and data of adversaries, we can estimate the probability of threats, as shown in Table 5.1.

According to the identified threats, the attractiveness of assets to each threat should be determined to obtain the possibility of attack scenarios. The attractiveness of an asset may change with different threats. Consequently, the Expert should assess the attractiveness of each asset to each threat from the perspective of adversaries. According to various threats, one or more following factors [15] may be considered: (i) potential for mass casualties/fatalities; (ii) extensive property damage; (iii) proximity to national assets or landmarks; (iv) possible disruption or damage to critical infrastructure; (v) disruption of the national, regional, or local economy; (vi) ease of access to target; (vii) media attention or possible interest of the media; (viii) company reputation and brand exposure; (ix) vulnerability of installations exposed to attacks; (x) Potential for triggering domino effects. The potential for triggering domino effects is the main difference in attractiveness assessment between the chemical industry and other sectors without escalation effects. The domino effect potential assessment is a simple assessment of domino effects, and simple domino effect models are recommended, such as graph metrics [47]. According to graph theory, closeness metrics measure the centrality of a node in a network [48]. The out-closeness metric can reflect installations' potential contribution to the escalation of domino effect, while the in-closeness metric represents the vulnerability of installations to get damaged during domino effects [43]. The installation with a high out-closeness value has a

Table 5.1 SRA methodology for threat assessment (*Note* Adapted from API [15])

Threat level	Description
Very low	Indicates little or no credible evidence of capability or intent and no history of actual or planned threats against the asset or similar assets (e.g., "no expected attack in the life of the facility's operation")
Low	Indicates that there is a low threat against the asset or similar assets and that few known adversaries would pose a threat to the asset (e.g., "1 event or more is possible in the life of the facility's operation")
Medium	Indicates that there is a possible threat to the asset or similar assets based on the threat's desire to compromise similar assets. Still, no specific threat exists for the facility or asset (e.g., "1 event or more in 10 years of the facility's operation")
High	Indicates that a credible threat exists against the asset or similar assets based on knowledge of the threat's capability and intent to attack the asset or similar assets, and some indication exists of the threat specific to the company, facility, or asset (e.g., "1 event or more in 5 years of the facility's operation")
Very high	Indicates that a credible threat exists against the asset or similar assets; that the threat demonstrates the capability and intent to launch an attack; that the subject asset or similar assets are targeted or attacked on a frequently recurring basis; and that the frequency of an attack over the life of the asset is very high (e.g., "1 event per year")

high potential to start domino effects, while the installation with a high in-closeness value is likely to be damaged by domino effects caused by other installations. Let's assume that adversaries expect to trigger a domino effect to induce severe consequences or indirectly damage installations with high in-closeness values. In that case, the installations with a high out-closeness value may be the targets.

Based on the above attractiveness analysis, the attractiveness of an asset to a threat represented by the likelihood that an installation is attacked by a threat can be determined based on the attractiveness levels shown in Table 5.2.

Table 5.2 Attractiveness evaluation table (API [15])

Attractiveness level	Likelihood L_A	Description
Very low	$0.0 \leq L_A < 0.2$	The threat would have little to no level of interest in the asset
Low	$0.2 \leq L_A < 0.4$	The threat would have some degree of interest in the asset, but it is not likely to be of interest compared to other assets
Medium	$0.4 \leq L_A < 0.6$	The threat would have a moderate degree of interest in the asset relative to other assets
High	$0.6 \leq L_A < 0.8$	The threat would have a high degree of interest in the asset relative to other assets
Very high	$0.8 \leq L_A \leq 1$	The threat would have a very high degree of interest in the asset, and it is a preferred choice relative to other assets

5.4.3 Vulnerability Assessment of Installations Against Direct and Threats

In this step, the vulnerability assessment aims to identify possible primary scenarios that may trigger domino effects and thus determine the probability of the primary scenarios caused by threats or hazards. The widely used risk assessment methods such as fault tree, event tree, and Bayesian network may be used to analyze possible primary scenarios caused by hazards and threats. In terms of the vulnerability of installations exposed to intentional attacks, it can be any weakness that may be exploited by an attacker to gain access to direct targets and to successfully execute an attack [15]. Before identifying primary scenarios caused by intentional attacks, we need to assess the likelihood that an attack is successfully carried out. In this chapter, the EASI model [42] is recommended to obtain the conditional probability that an attack is successfully carried out. According to the model, the likelihood depends on three parameters:

(1) The conditional probability that detected information is communicated to the emergency force. It depends on the training in communication equipment, maintenance, dead spot in radio communication, and the stress experienced during actual attacks [42].
(2) The conditional probability that an attack is detected on time. It depends on the attack path, detection measures along the path, and emergency response time. If the needed time for an attacker to pass the path between the detection position and the attack target is larger than the emergency response time, the attack is assumed to be interrupted. Otherwise, the attacker would reach the location of the target. To improve the likelihood of interrupting intentional attacks, the time the attacker is detected should be as early as possible. Besides, multiple detection measures may be implemented to increase the detection probability. Furthermore, delay measures may be used to delay the time that the attacker reaches the target and thus provide more time for the response of emergency force.
(3) The conditional probability that the attack is successfully executed when the adversary gets access to the target. It is determined by the performance that the attacker correctly uses the weapon and the reliability of the weapon [49].

5.4.4 Assessment of the Evolution of Domino Effects

Based on the vulnerability assessment method illustrated in Sect. 5.4.2, we can obtain all the primary scenarios that may induce domino effects. These primary scenarios can be intentional or unintentional. Besides, multiple failures may be present in one primary scenario. According to the primary scenarios, the graph-based models presented in Chaps. 2, 3 and 4 can be selected to analyze the possible subsequent domino effects and to calculate the failure probability of installations and the death

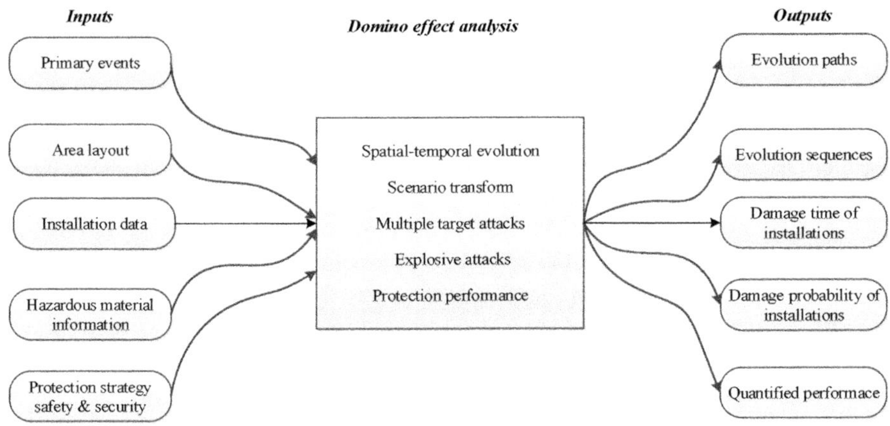

Fig. 5.5 The framework for domino effect analysis

probability of humans. The inputs and outputs of the domino effect analysis are shown in Fig. 5.5.

5.4.5 Consequence Analysis

Domino effects may result in many consequences, such as fatalities, economic losses, and environmental damage. Based on the consequence analysis method proposed by API [15], five categories of consequences are considered: (i) fatalities and injuries, (ii) property damage, (iii) environmental impacts, (iv) business interruption, (v) damage to reputation or negative publicity.

Loss of human life

Humans are vulnerable to toxic gas, heat radiation, overpressure, and fragments. Since the probability of humans dying from fragments is much lower than dying from the other harmful effects, we focus on the latter effects. In this chapter, therefore, the probit models for human vulnerability developed by TNO [50] are recommended to obtain the death probability caused by hazardous effects. For example, the acute intoxication of humans caused by a toxic gas is a function of the harmful effects of the toxic gas, the toxic gas concentration, and the exposure time. In terms of the death caused by overpressure, the vulnerability of humans exposed to heat radiation is also a dynamic process and depends on the exposure time, and the received heat radiation. For the damage caused by heat radiation, the superimposed effects may be considered since the heat radiation received by humans may vary with the number of hazardous installations being on fire during the spatial–temporal evolution of hazards.

Property damage

The property damage caused by domino effects in a chemical plant should account for all possible damaged installations based on the results obtained from the vulnerability assessment and domino effect analysis. The property damage is always monetarized as the cost of lost property [51].

Business loss

The business loss is related to the operation interruption of the chemical plant caused by domino effects, and the interruption time can be used as an indicator of business loss [15].

Environmental damage

Besides the damage to humans, property, and business, environmental impacts are not neglected. Spills of hazardous materials may cause damage to the environment, indirectly affecting the population, plants, and animals by contaminating land, surface water, and groundwater [52]. A certain period of time is needed for the recovery of the environment. As a result, the damage to the environment may be assessed by the time needed to recover [15].

Reputation or negative publicity

The damage to reputation or negative publicity is challenging to evaluate. Some indirect indicators may be used to determine it, such as the degree of loss of reputation or business viability, the attention of regulatory agencies, and the report of media [15].

According to the above analysis, a consequence evaluation table can be obtained, as shown in Table 5.3.

For instance, the damage to two above-ground gasoline tanks in a decked area can lead to on-site injuries that are not widespread but only in the vicinity of the incident location (low); over €1 billion to €10 million loss in property damage (medium); minor environmental impacts to immediate incident site area only, less than 1 year to recover (low); Short-term (>2 weeks to 3 months) business interruption/expense (low); and a medium loss of reputation or business viability (medium). Since two types of consequences are in the medium level, the total consequence level is medium.

5.4.6 *Risk Evaluation*

To determine whether measures need to be taken to reduce domino effect risks, risk evaluation should be conducted based on the likelihood and consequence of domino effects. To make it more user-friendly, a risk matrix is recommended, as shown in Fig. 5.6. According to the risk attitude of decision-makers, the risk threshold

Table 5.3 Consequence evaluation table (API [15])

Consequence level	Description (already look at the worst-case parameters)
Very low	(i) Possibility of minor injury on-site; no fatalities or injuries anticipated off-site (ii) Up to €X loss in property damage (iii) No environmental impacts (iv) Very short-term (up to X weeks) business interruption/expense (v) Very low or no impact or loss of reputation or business viability; mentioned in the local press
Low	(i) On-site injuries that are not widespread but only in the vicinity of the incident location; no fatalities or injuries anticipated off-site (ii) €X to €X loss in property damage (iii) Minor environmental impacts to immediate incident site area only, less than X year(s) to recover (iv) Short-term (>X week to Y months) business interruption/expense (v) Low loss of reputation or business viability; query by the regulatory agency; significant local press coverage
Medium	(i) Possibility of widespread on-site serious injuries; no fatalities or injuries anticipated off-site (ii) Over €X to €X loss in property damage (iii) Environmental impact on-site and/or minor off-site impact, Y year(s) to recover (iv) Medium-term (Y to Z months) business interruption/expense (v) Medium loss of reputation or business viability; attention of regulatory agencies; national press coverage
High	(i) Possibility of X to Y on-site fatalities; the possibility of off-site injuries (ii) Over €X to €X loss in property damage (iii) Tremendous environmental impact on-site and/or large off-site impact, between Y and Z years to recover (iv) Long-term (X to Y years) business interruption/expense (v) High loss of reputation or business viability; prosecution by the regulator; extensive national press coverage
Very high	(i) Possibility of any off-site fatalities from large-scale toxic or flammable release; possibility of multiple on-site fatalities (ii) Over €X loss in property damage (iii) Major environmental impact on-site and/or off-site (e.g., large-scale toxic contamination of public waterway), more than XX years/poor chance of recovery (iv) Very long-term (>X years) business interruption/expense; large-scale disruption to the national economy, public or private operations; loss of critical data (v) Very high loss of reputation or business viability; international press coverage

* X, Y, Z, and XX are variables that users should determine

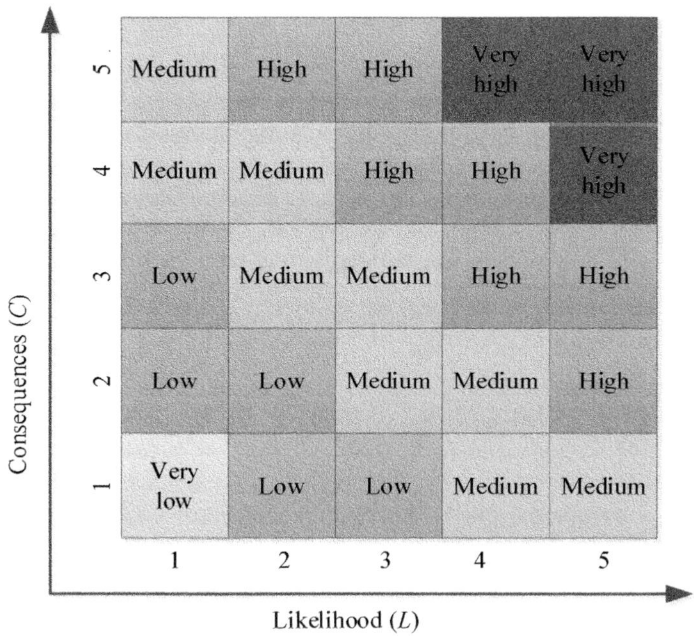

Fig. 5.6 Risk matrix for domino effect risk evaluation (to be filled in based on Tables 5.2 and 5.3)

may differ. If the obtained risk is lower than the threshold, the risk is acceptable; otherwise, add-on risk reduction measures may be taken to make the risk lower than the threshold.

5.4.7 Risk Treatment

Risk treatment is an essential step in the integrated management framework, focusing on selecting and implementing appropriate measures to deal with domino effect risk. The common-used risk treatment strategies include risk retention, risk avoidance, risk reduction, risk transfer [51]. One or more risk treatment strategies may be selected according to distinctive risk evaluation results and the preferences of decision-makers.

If the domino effect risk of a chemical plant is lower than the risk tolerability criteria, a risk retention strategy may be adopted. Based on the risk retention strategy, the chemical plant will accept the domino effect risk and be prepared if a domino effect occurs. If the domino effect risk of a chemical plant exceeds the tolerability threshold, the risk is unacceptable and risk treatment strategies such as risk avoidance, risk reduction, and risk transfer may be taken to make the risk lower than the threshold. The following risk treatment possibilities exist:

(1) **Risk avoidance**

In terms of domino effect risks, risk avoidance may refer to the inherent safety and security measures. For instance, using less hazardous or attractive materials and processes can reduce or eliminate hazards and threats and thus decreases the likelihood of primary scenarios. Besides, changing the design of plant layout (e.g., providing ample layout spacing) or operations (e.g., reducing the hazardous material unloading velocity) may limit the hazardous effects of escalation so as to lower the escalation potential. Other inherent safety and security measures are shown in Sect. 5.2.1.

(2) **Risk reduction**

Risk reduction can be needed and it is determined by additional protection measures, decision in ideally influenced by the performance of the used protection measures [44]. For example, we can install additional detection sensors to improve the detection probability of abnormal events (unintentional events and intentional events) and provide more time for the initiation of emergency response, thus decreasing the domino effect risk. Besides, delay measures such as fireproof coatings can be used to prevent fire-induced escalation, and fences may be used to delay intentional attacks originating from outside the chemical plant. Furthermore, emergency response forces may be improved to shorten the time needed to respond to an abnormal event and enhance the capability to control adversaries and accident scenarios. Besides the risk reduction performance of protection measures, the costs of protection measures may also be considered in the decision-making on risk protection strategies since there is always a budget for the investment of safety and security. After implementing protection measures, a risk assessment should be conducted again to check whether the reduced risk is satisfactory.

(3) **Risk transfer**

Risk transfer aims to shift the risk to another party. Insurance is a typical risk transfer measure in which a specified risk is transferred from a chemical plant to the insurer by purchasing insurances [53]. In that case, if a domino effect event occurs, the insurance company pays the cost. However, the risk is not reduced by the insurance. Usually, the company can only get an economic income from the insurance company to compensate for a name of different losses. In terms of security risk, risk transfer may occur without any intentional actions. For example, a chemical plant invests in protection measures to prevent intentional attacks while the nearby chemical plants in the chemical cluster do not take any protection measures. In that case, the adversary of the chemical cluster may aim to attack the chemical companies without protection measures, resulting in a risk transfer from the plant with protection measures to the plant without protection measures.

5.5 Conclusions

In this chapter, an integrated domino effect management framework is developed to tackle the possible domino effects caused by intentional threats or hazards. According to this framework, safety and security measures should be integrated to avoid possible overlaps. Besides, both intentional threats and unintentional hazards should be identified to prevent underestimating the domino effect risk. The vulnerability assessment is divided into two parts: the vulnerability of installations directly against hazards and attacks and the vulnerability of installations exposed to possible domino effects. Consequence analysis accounts for fatalities, property loss, environmental impacts, business loss, and reputation or negative publicity. Finally, the domino effect risk can be calculated and evaluated to determine whether the risk is acceptable. If the risk is unacceptable, additional protection measures should be taken until the risk is lower than the pre-defined risk tolerability criteria. Based on the framework, decision-making approaches such as cost-effectiveness analysis and cost–benefit analysis can be developed to obtain optimal protection strategies.

References

1. Chen C, Reniers G, Khakzad N (2020) Cost-benefit management of intentional domino effects in chemical industrial areas. Process Saf Environ Prot 134:392–405
2. Chen C (2021) A dynamic and integrated approach for modeling and managing domino-effects. Delft University of Technology
3. Landucci G, Argenti F, Spadoni G, Cozzani V (2016) Domino effect frequency assessment: the role of safety barriers. J Loss Prev Process Ind 44:706–717
4. Janssens J, Talarico L, Reniers G, Sörensen K (2015) A decision model to allocate protective safety barriers and mitigate domino effects. Reliab Eng Syst Saf 143:44–52
5. Ghasemi AM, Nourai F (2017) A framework for minimizing domino effect through optimum spacing of storage tanks to serve in land use planning risk assessments. Saf Sci 97:20–26
6. Khakzad N, Reniers G (2017) Cost-effective allocation of safety measures in chemical plants w.r.t land-use planning. Saf Sci 97:2–9
7. Reniers G, Khakzad N, Gelder PV (2017) Security risk assessment: in the chemical and process industry. De Gruyter, Berlin
8. Baybutt P (2002) Assessing risks from threats to process plants: threat and vulnerability analysis. Process Saf Prog 21(4):269–275
9. Baybutt P (2003) Strategies for protecting process plants against terrorism, sabotage and other criminal acts. Homeland Defence J 2(1)
10. Bier VM, Nagaraj A, Abhichandani V (2005) Protection of simple series and parallel systems with components of different values. Reliab Eng Syst Saf 87(3):315–323
11. Reniers G, Dullaert W, Audenaert A, Ale BJ, Soudan K (2008) Managing domino effect-related security of industrial areas. J Loss Prev Process Ind 21(3):336–343
12. Lee Y, Kim J, Kim J, Kim J, Moon I (2010) Development of a risk assessment program for chemical terrorism. Korean J Chem Eng 27(2):399–408
13. CCPS (2003) Guidelines for analyzing and managing the security vulnerabilities of fixed chemical sites. American Institute of Chemical Engineers
14. American Petroleum Institute (API) (2004) Recommended practice for security vulnerability assessment for petroleum and petrochemical facilities. American Petroleum Institute

15. API (2013) ANSI/API Standard 780—security risk assessment methodology for the petroleum and petrochemical industry. American Petroleum Institute
16. Baybutt P (2017) Issues for security risk assessment in the process industries. J Loss Prev Process Ind 49:509–518
17. Paté-Cornell E, Guikema S (2002) Probabilistic modeling of terrorist threats: a systems analysis approach to setting priorities among countermeasures. Mil Oper Res 7(4):5–23
18. Staalduinen MAv, Khan F, Gadag V (2016) SVAPP methodology: a predictive security vulnerability assessment modeling method. J Loss Prevent Process Indus 43:397–413
19. Moore DA, Fuller B, Hazzan M, Jones JW (2007) Development of a security vulnerability assessment process for the RAMCAP chemical sector. J Hazard Mater 142(3):689–694
20. Matteini A, Argenti F, Salzano E, Cozzani V (2018) A comparative analysis of security risk assessment methodologies for the chemical industry. Reliab Eng Syst Saf
21. Cox JLA (2009) Game theory and risk analysis. Risk Anal 29(8):1062–1068
22. Cox LA Jr (2008) Some limitations of "Risk = Threat x Vulnerability x Consequence" for risk analysis of terrorist attacks. Risk Anal 28(6):1749–1761
23. Powell R (2007) Defending against terrorist attacks with limited resources. Am Politi Sci Rev 101(3):527–541
24. Reniers G, Soudan K (2010) A game-theoretical approach for reciprocal security-related prevention investment decisions. Reliab Eng Syst Saf 95(1):1–9
25. Rios J, Insua DR (2012) Adversarial risk analysis for counterterrorism modeling. Risk Anal 32(5):894–915
26. Tambe M (2011) Security and game theory: algorithms, deployed systems, lessons learned. Cambridge University Press
27. Liu D, Wang X, Camp J (2008) Game-theoretic modeling and analysis of insider threats. Int J Crit Infrastruct Prot 1:75–80
28. Reniers GLL, Audenaert A (2014) Preparing for major terrorist attacks against chemical clusters: Intelligently planning protection measures w.r.t. domino effects. Process Saf Environ Protect 92(6):583–589
29. Kletz TA (2003) Inherently safer design—its scope and future. Process Saf Environ Prot 81(6):401–405
30. Kletz TA, Amyotte P (2010) Process plants: a handbook for inherently safer design. CRC Press
31. Khan FI, Amyotte PR (2003) How to make inherent safety practice a reality. Canad J Chem Eng 81(1):2–16
32. CCPS (2001) Layer of protection analysis: simplified process risk assessment. AIChE–CCPS, New York
33. CCPS (2011) Layer of protection analysis: simplified process risk assessment. American Institute of Chemical Engineers—Center of Chemical Process Safety, New York
34. Reniers G, Van Lerberghe P, Van Gulijk C (2015) Security risk assessment and protection in the chemical and process industry. Process Saf Prog 34(1):72–83
35. Reniers G, Landucci G, Khakzad N (2020) What safety models and principles can be adapted and used in security science? J Loss Prevent Process Indus 64
36. De Dianous V, Fievez C (2006) ARAMIS project: a more explicit demonstration of risk control through the use of bow–tie diagrams and the evaluation of safety barrier performance. J Hazard Mater 130(3):220–233
37. Landucci G, Argenti F, Tugnoli A, Cozzani V (2015) Quantitative assessment of safety barrier performance in the prevention of domino scenarios triggered by fire. Reliab Eng Syst Saf 143:30–43
38. Khakzad N, Landucci G, Reniers G (2017) Application of dynamic Bayesian network to performance assessment of fire protection systems during domino effects. Reliab Eng Syst Saf 167:232–247
39. Khakzad N, Landucci G, Cozzani V, Reniers G, Pasman H (2018) Cost-effective fire protection of chemical plants against domino effects. Reliab Eng Syst Saf 169:412–421
40. Zhou J, Reniers G, Khakzad N (2016) Application of event sequence diagram to evaluate emergency response actions during fire-induced domino effects. Reliab Eng Syst Saf 150:202–209

41. Zhou J, Reniers G (2018) Petri-net based evaluation of emergency response actions for preventing domino effects triggered by fire. J Loss Prev Process Ind 51:94–101
42. Garcia ML (2007) Design and evaluation of physical protection systems. Elsevier
43. Chen C, Reniers G, Khakzad N (2019) Integrating safety and security resources to protect chemical industrial parks from man-made domino effects: a dynamic graph approach. Reliab Eng Syst Saf 191
44. Chen C, Reniers G, Khakzad N (2020) A thorough classification and discussion of approaches for modeling and managing domino effects in the process industries. Saf Sci 125
45. Chen C, Li C, Reniers G, Yang F (2021) Safety and security of oil and gas pipeline transportation: a systematic analysis of research trends and future needs using WoS. J Cleaner Product 279
46. CCPS (1992) Guidelines for hazard evaluation procedures, 2nd edn. American Institute of Chemical Engineers, New York
47. Khakzad N, Landucci G, Reniers G (2017) Application of graph theory to cost-effective fire protection of chemical plants during domino effects. Risk Anal 37(9):1652–1667
48. Freeman LC (1978) Centrality in social networks conceptual clarification 1(3):215–239
49. Stewart MG, Mueller J (2012) Terror, security, and money: balancing the risks, benefits, and costs of critical infrastructure protection, pp 513–533
50. Van Den Bosh C, Merx W, Jansen C, De Weger D, Reuzel P, Leeuwen D, Blom-Bruggerman J (1989) Methods for the calculation of possible damage (Green Book). Committee for the Prevention of Disasters, The Hague, NL
51. Reniers GL, Van Erp HN (2016) Operational safety economics: a practical approach focused on the chemical and process industries. Wiley
52. Bonvicini S, Antonioni G, Cozzani V (2018) Assessment of the risk related to environmental damage following major accidents in onshore pipelines. J Loss Prev Process Ind 56:505–516
53. Abrahamsen EB, Asche F (2010) The insurance market's influence on investments in safety measures. Saf Sci 48(10):1279–1285

Chapter 6
An Economic Approach for Domino Effect Management

6.1 Introduction

Domino effects are responsible for many catastrophic disasters in the chemical and process industry,[1] such as the Buncefield accident in 2005 [3] and the Jaipur disaster in 2009 [4]. Both unintentional (safety-related) events (e.g., mechanical failure, human error, and natural disasters) and intentional (security-related) events (e.g., terrorist attacks) can lead to a domino effect.

In light of possible intentional and unintentional escalation events, Reniers and Soudan [5] proposed setting up a Multi-Plant Council (MPC) to stimulate the prevention cooperation, thus preventing and mitigating domino effects from a chemical cluster-level perspective. Reniers and Audenaert [6] provided a systematic method to reduce the potential consequences of domino effects triggered by intentional attacks. Janssens et al. [7] developed a mathematical model to optimize the allocation of safety barriers for mitigating domino effects. Landucci et al. [8] developed a model to assess the vulnerability of industrial installations and possible escalation effects triggered by homemade explosives. Zhou and Reniers [9] optimized emergency response strategies for tackling domino effects caused by multiple simultaneous fires. Hosseinnia et al. [10] proposed a matrix for decision-making on emergency response. Khakzad and Reniers [11] proposed to make some of the storage tanks empty to prevent the escalation of domino effects. Although the previous work for domino effect management has made significant progress in preventing and mitigating domino effects, economic issues are very often, if not always neglected in decision-making. Economic aspects of safety and security nonetheless play an indispensable role in the decision-making of protection strategies because companies usually have a budget for safety and security. Besides, the limited safety and security resources need to be optimized since resources allocated to one target are not available for others, which is referred to in economic literature as the "opportunity cost"

[1] This chapter is mainly based on two publications: Chen et al. [1], Chen et al. [2].

C. Chen et al., *Integrating Safety and Security Management to Protect Chemical Industrial Areas from Domino Effects*, Springer Series in Reliability Engineering,
https://doi.org/10.1007/978-3-030-88911-1_6

[12–14]. In other words, economic issues greatly impact the effectiveness of prevention strategies and the profitability of a company in the long term [15]. Therefore, operational safety economics needs to be applied in decision-making on safety and security strategies [15]. Safety economics may be defined as a transdisciplinary and interdisciplinary field of academic research focusing on the interdependencies and coevolution of economics and safety for the trade-off between safety and economics [1]. Safety economics aims to support decision-making on safety investments to make decisions more profitable under the premise of safety criteria. It may be interpreted as balancing the costs of risks and economic benefits or balancing the costs for decreasing risks and safety benefits.

However, there is a research gap between safety economics and domino effect management due to the complexity and uncertainty of domino effect evolution, the monetization of protection costs and benefits, and insufficient information about the intelligent and strategic adversaries. Thus, this chapter develops a cost–benefit approach based on safety economics to manage domino effects in the chemical and process industry. First, safety economics is introduced in Sect. 6.2. Based on safety economics, a commonly used approach (cost–benefit analysis) is selected as a decision-making tool for managing domino effects. A cost–benefit management approach is developed in Sect. 6.3, and an optimization algorithm is presented in Sect. 6.4. The conclusions are drawn in Sect. 6.5.

6.2 Safety Economics

Safety economics can be traced back to the 1960s when economics was considered in safety [16]. From an economic perspective, safety is regarded as a resource-absorbing product or service. The objective of safety economics is to produce safety in a suitable amount and as cheaply as possible [16]. Safety economics is used to obtain a balance between safety and the cost of safety measures as well as develop an optimal combination of rewards and penalties to balance the costs of safety measures under the framework of competition and bargaining [16]. Therefore, an optimal safety strategy should minimize the sum of potential accident costs and the accident prevention costs rather than reduce the frequency and consequences of accidents to be as low as possible [17]. Besides, operational safety economics can be regarded as a decision-making tool for the safety and security of organizations [15]. According to past research, the contributions of economics on safety may be divided into three categories: (i) identifying and measuring the costs of accidents, (ii) understanding the relationship between business and safety, and (iii) achieving a trade-off between safety and other goals of an organization. Besides the economic perspective of safety, safety should also be considered in economic activities since safety risk plays a vital role in developing a sustainable economic organization [15, 18]. In terms of the feasibility of a new economic project or product, safety can be a non-negligible factor. For instance, the paradigm "safety at all costs" was proposed for nuclear plant safety in the 1980s, but it may increase questionable back fitting measures and production costs [19, 20]. As a result, safety economics may be defined as:

> A transdisciplinary and interdisciplinary field of academic research focusing on the interdependencies and coevolution of economics and safety for the decision-making between safety and economics [1].

Some approaches used in safety economics can be discussed such as risk-based optimization, minimal total safety cost approach, cost–benefit analysis, cost-effectiveness analysis, multi-objective optimization, and the game-theoretical approach.

6.2.1 *Risk-Based Optimization*

Risk-based optimization is a widely used approach in safety economics, considering both the costs of safety measures and risk tolerability criteria. This approach consists of three steps: risk assessment, risk management, and economic analysis, as shown in Fig. 6.1. According to this approach, the first step is to describe the system (e.g., individual and industrial company). Besides, risk acceptance criteria are determined

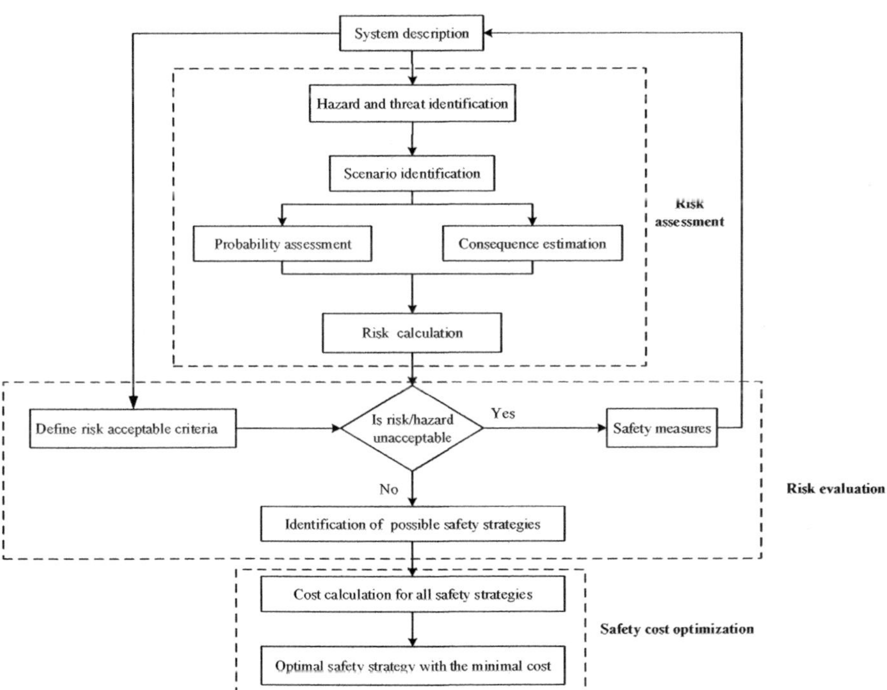

Fig. 6.1 Flowchart of risk-based optimization (Chen et al. [1])

and used to judge whether the risk of the system is acceptable. If the risk is unacceptable, additional protection measures should be implemented to make the risk following the criteria. The second step is hazard identification which is used to identify possible hazards. Many hazard identification methods are available in literature, such as HAZOP, LOPA, Checklists, and What-If analysis [21]. In terms of security risk, possible threats should also be identified in this step. Following the hazard and threat identification, possible undesired scenarios caused by these hazards and threats can be determined in the third step. Next, probabilities and consequences of each scenario need to be estimated. Some analysis methods can support the probability calculation, such as fault tree analysis, Bayesian network, and Monte Carlo simulation. Risk can be regarded as a function of the likelihood and consequence of undesired scenarios. Following the likelihood estimation and consequence analysis, the risk can be calculated. Subsequently, the risk should be evaluated based on the selected risk tolerability criteria.

The risk tolerability criteria may vary with different systems, decision-makers, or jurisdictions and may be determined by experience, standards, laws, etc. [22]. As a result, different risk criteria are available for decision-makers. For example, individual risk or societal risk criteria may be used if decision-makers are concerned more about possible fatalities than other consequences. Individual risk represents the likelihood of death if an individual is exposed to hazards. In contrast, societal risk means the cumulative probability of N fatalities (using F-N curves) given a population (a group of people) is exposed to risks [23]. Besides the individual and societal risk criteria, risk-based economic losses and environmental damage can also be applied [24]. For each risk criterion, a risk threshold may be defined to determine if the risk is acceptable. If the calculated risk is below the threshold, it is acceptable. If the calculated risk is higher than the threshold, risk reduction measures should be taken to decrease the risk under the condition of the maximum protection budget. If required protection measures are neither available nor feasible (cost-wise, operation-wise, etc.), the system may need to be discontinued.

Safety cost optimization aims to determine the optimal safety strategy with minimal costs from all the possible protection strategies obtained in a risk evaluation. The main task of this step is to calculate the costs of protection strategies. According to previous studies [15, 25], the costs may be divided into two types: direct cost and indirect cost.

The direct cost represents the investment for implementing protection measures, involving initial cost and recurring cost. The initial cost is a one-off expense that only occurs at the initial stage before implementation of the safety measures. It can be divided into two categories: initiation cost (e.g., material and design) and installation cost (e.g., labor cost and equipment cost). The recurring cost represents the ongoing expenses for operating the protection measures, including operation cost, maintenance cost, and inspection cost, etc. The initial cost and the recurring safety cost should not be directly summed since they do not simultaneously occur. To deal with this problem, a discount rate can convert the initial cost to an equivalent annual cost (EAC) or discount the recurring cost to the net present value (NPV).

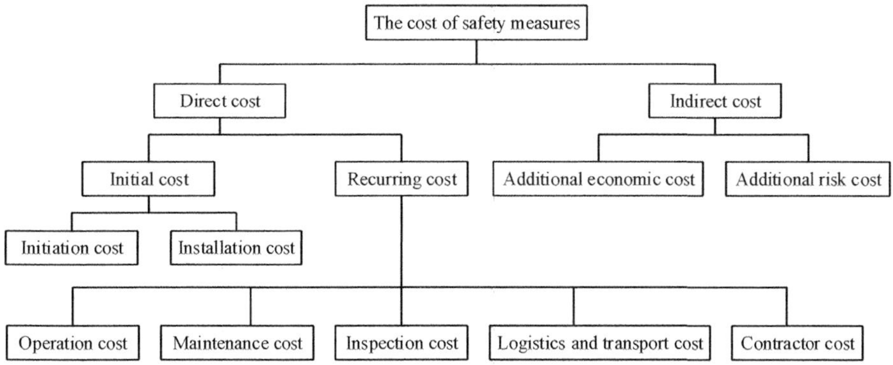

Fig. 6.2 The components of the costs of safety measures (Chen et al. [1])

The indirect cost refers to the additional economic loss and additional risk caused by protection measures. It may play a leading role in protection costs. For example, the additional economic cost (e.g., economic recession and high unemployment rate) of protection measures (e.g., social distance and face mask) for preventing Covid-19 is much higher than the direct costs for implementing these measures. Besides, these protection measures may also bring additional risks. For instance, taking a face mask may increase transportation accidents due to foggy glasses caused by the mask. In terms of a specific case, some types of costs listed in Fig. 6.2 may be neglected if they are much lower than others. According to the costs of all protection strategies, the optimal protection strategy that has the minimum cost can be obtained.

6.2.2 Minimum Total Cost Approach

In the risk-based approaches, risk acceptance criteria are used to select viable protection strategies that can make the actual risk lower than a risk threshold. Cost analysis is used to obtain the optimal protection strategy with the minimum implementation cost. In the minimum cost approach, not only the implementation cost but also the potential accident cost are considered [26]. This approach is to obtain the optimal safety strategy with the minimum total cost (the sum of the implementation cost and the accident cost). As shown in Fig. 6.3, this approach also involves hazard identification, scenario identification, probability assessment, and consequence estimation. Based on the above steps, a cost analysis can be conducted to obtain both the potential accident costs and the implementation cost of protection strategies. The total cost is obtained by summing the implementation cost of the protection strategy and the expected value of potential accident cost (based on the calculated risk) after implementing the protection strategy. As a result, the protection strategy with the minimum total cost should be selected as the optimal strategy.

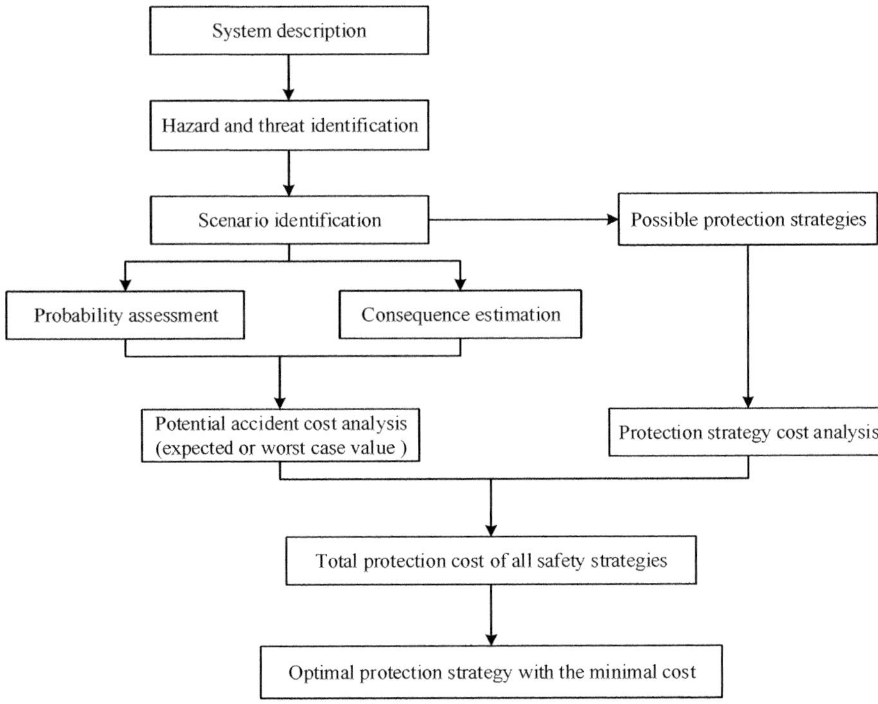

Fig. 6.3 Flowchart of the minimal total safety cost approach (Chen et al. [1])

The cost analysis for implementing a protection strategy in this approach is identical to that in the risk criteria-based approach (see Sect. 6.2.1). Thus the calculation and monetization of accident costs is the most important task in the minimum total cost approach. Many accident consequence cost calculation methods are available in literature. For instance, Sun et al. [27] divided accident costs into two categories: insured costs and uninsured costs. Besides, Gavious et al. [28] demonstrated that accident costs should consist of four sub-categories: direct costs, indirect costs, payment (the increased payment to employees), and immeasurable costs. Moreover, Reniers and Van Erp [15] divided accident costs into ten categories, as shown in Fig. 6.4.

Among the ten categories of accident costs, the cost of human loss has obtained the most attention since the value of human life is an ethical issue and thus not so easy to monetize. Since it is difficult to directly obtain the monetary value of human life, it can be indirectly monetized via those that can be directly quantified. For example, the willingness to pay (WTP) method uses the money that a company is willing to pay to reduce the risk of human loss to indirectly obtain the value of human life. Besides, the WTPs of a group of humans can be used to calculate the average individual value called the Value of a Statistical Life (VSL) [29]. The VSL is a location-based value that depends on the region (varies in 1–10 million Euros) and that may vary with

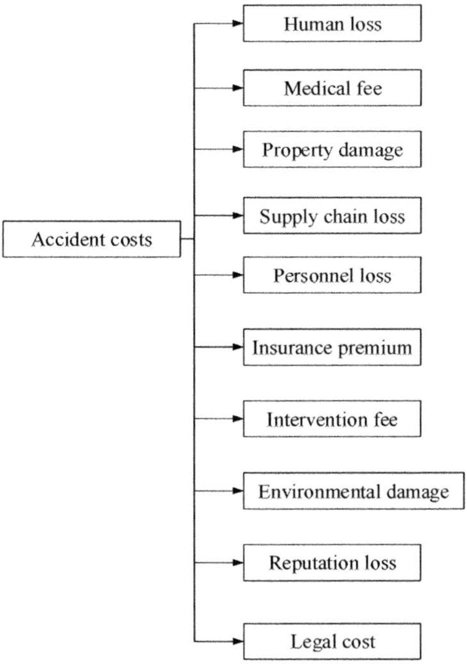

Fig. 6.4 The components of accident costs (Chen et al. [1])

time [30]. Other widely used methods include the human capital method and the Quality-Adjusted Life Years (QALY) [31].

6.2.3 Cost–Benefit Analysis

Cost–benefit analysis (CBA) is a widely used approach for decision-making on prevention strategies [13]. Figure 6.5 shows the procedure of a cost–benefit analysis approach for decision-making about protection strategies.

In a cost–benefit analysis, costs and benefits are usually represented by money to compare different options [32]. In the safety, security, and health domain, it is used to value costs and benefits and enable a broad comparison among difficult risk reduction measures, obtaining the optimal protection strategies. As a result, all costs related to the implementation of an intervention must be identified and monetarized. Similarly, all relevant benefits (usually avoided costs) caused by the intervention need to be identified and represented in monetary terms. The implementation cost of protection measures may be regarded as a sunk cost for avoiding and mitigating potential costs of undesired events. The avoided costs thus can be considered as the expected benefits of the investment in protection measures. To calculate the expected benefit of protection investments, a baseline protection strategy needs to be defined, such as the strategy "without any additional protection investment" (the so-called

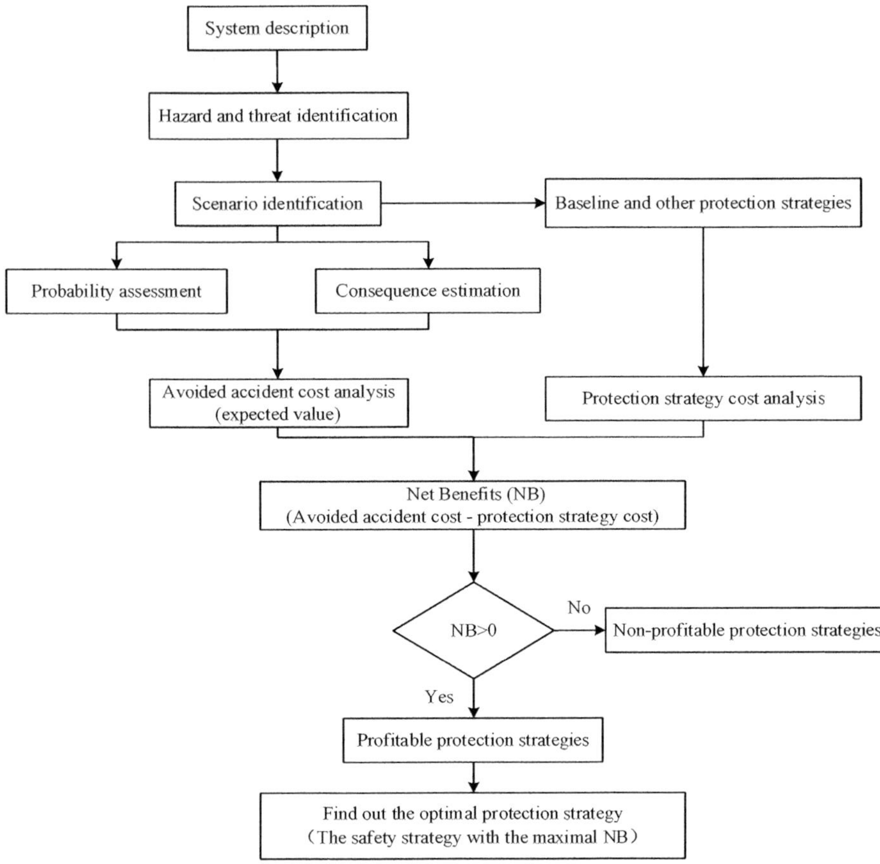

Fig. 6.5 Flowchart of cost–benefit analysis approach (Chen et al. [1])

"naked option"). As a result, the hypothetical benefit of a protection strategy can be expressed as the difference between the undesired event cost of the baseline strategy minus the undesired event cost of the protection. The net benefit of a protection strategy can be easily represented as the difference between the expected benefit of a protection strategy and the equivalent annual cost of the protection strategy. A protection strategy is profitable and may be adopted if the net benefit is larger than zero; otherwise, the protection strategy is non-profitable and may be abandoned if more profitable strategies are available. In other words, the protection strategy that has the maximum net benefit is the optimal protection strategy.

6.2.4 Cost-Effectiveness Analysis

Cost-effectiveness analysis (CEA) is a decision-making tool in which the costs and consequences of all the options are considered systematically [33]. CEA is always used in healthcare to assess specific interventions dominated by studies of prospective new interventions compared with current practice [34]. The estimation of protection costs is necessary for both the cost-effectiveness approach and the cost–benefit analysis approach. However, CBA is usually used to address those types of options that their performances can be measured by monetary values. In terms of CEA, the benefit is substituted with effectiveness analysis. In that case, CEA may be an alternative to overcome the problems associated with the monetarization of the costs related to undesired events. CEA is always used for decision-making between two different protection strategies or approving a new protection measure by estimating how much it costs to gain a unit of safety or security. Figure 6.6 illustrates a cost-effectiveness approach to determine whether a new strategy is adopted. Usually, an incremental cost-effectiveness ratio (ICER) [35] is defined as the ratio between the incremental cost (ΔC) and the corresponding incremental effectiveness (ΔE), The results of ICER can be divided into four categories: (i) If $\Delta C < 0$ and $\Delta E > 0$, the new strategy is dominant and should be recommended (less expensive and more effective); (ii) If $\Delta C > 0$ and $\Delta E < 0$, the current strategy is dominant and the new strategy (more

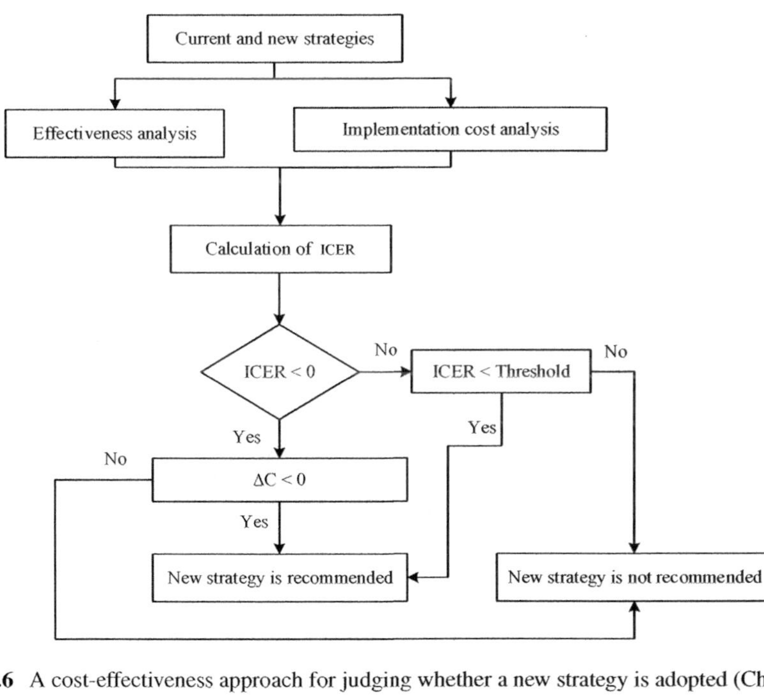

Fig. 6.6 A cost-effectiveness approach for judging whether a new strategy is adopted (Chen et al. [1])

expensive and less effective) should be rejected. (iii) If $\Delta C < 0$ and $\Delta E < 0$, it is a trade-off status in which the implementation of the new strategy can reduce implementation cost while decrease the effectiveness; (iv) If $\Delta C > 0$ and $\Delta E > 0$, it is also a trade-off status in which the implementation of the new strategy can improve effectiveness while needs more cost. In terms of the results of (iii) and (iv), a pre-determined threshold of ICER may be used to compare it with the actual ICER value. If the actual ICER is less than the threshold, the new strategy is recommended; otherwise, the new strategy is not cost-effective.

By applying the ICER, the units of protection cost and the protection outcome can be different from each other, making the CEA method more flexible than the CBA method. The protection outcome can be any indicator according to the decision-maker's preferences, such as fatalities, injuries, and the quality-adjusted life years (QALY).

6.2.5 Multi-objective Optimization

The decision-making on safety and security investment may involve multiple objectives, and the objective may be conflicting. For instance, the optimal safety and optimal economic benefit may not be simultaneously obtained since safety and economic benefits may be two conflicting objectives in the short run. As a result, multi-objective optimization may be an effective approach for balancing economic benefits and safety. Multi-objective optimization is a multiple-criteria decision-making approach to simultaneously optimize problems involving more than one objective function [36]. Multi-objective optimization has been used in inherent safety design to minimize the costs of a project and the corresponding risks [37] and minimize the operational costs in the entire life cycle [38]. In the design stage, the concept of inherently safer design can be used to reduce accident risk by using the principles such as substitution and simplification. However, the application of these principles may increase construction and operation costs. As a result, a multi-objective optimization model can be used in the design stage for optimizing the conflicting objectives: reduction of risk and construction and operation costs, as shown in Fig. 6.7.

The main task of multi-optimization is to identify and quantify the multi-objectives. As shown in Fig. 6.7, two objective functions can be obtained: (i) minimizing risk (or maximizing safety); (ii) minimizing safety cost. If we only consider one objective, a dominated solution can be obtained in which only one objective is optimal. By applying multi-objective approach and optimization algorithms (e.g., genetic algorithms and weighted sum methods), a set of Pareto solutions can be obtained. Pareto solutions refer to solutions in which there are no other solutions that can improve any objective without worsening one or more other objectives [11]. According to decision-makers' preferences, the final optimal strategy can be selected from a set of Pareto solutions. Because multi-objective optimization usually provides a set of Pareto strategies, the decision-makers need to selects the most preferred one.

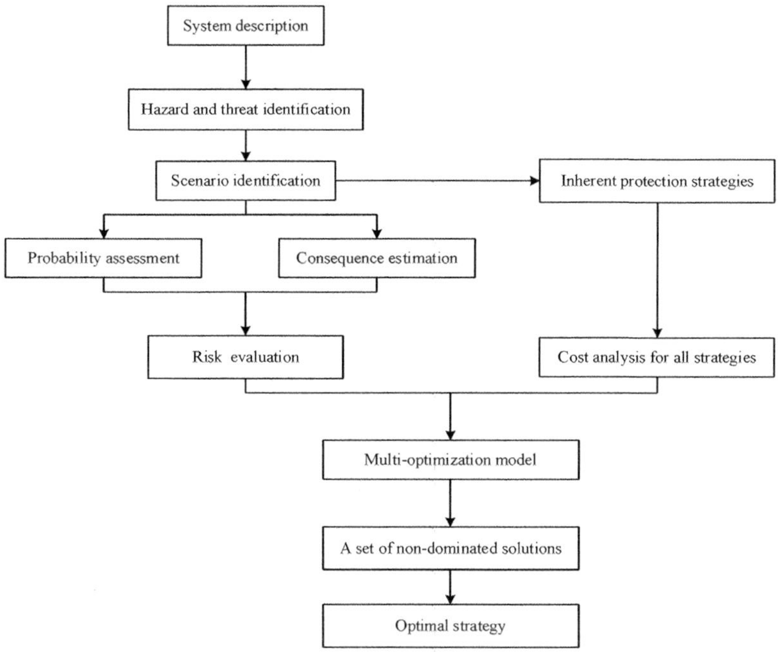

Fig. 6.7 A flowchart of a multi-optimization approach (Chen et al. [1])

Therefore, the decision-maker also plays an essential role in the optimization process, which is different from the above approaches that only have one solution.

6.2.6 *Game Theoretical Approach*

The preceding five methods are mainly used to deal with the decision-making problems involving only one single decision-maker. However, multiple decision-makers may be engaged in a protection investment. For instance, since the safety and security policy of a company always needs to follow the safety and security regulations, the decision-makers of the company are influenced by safety and security regulators [39]. The company may only be expected to maximize its economic benefits while the regulators aim to reduce the societal costs of undesired events. Besides, decision-making among different companies in a chemical cluster is also interdependent due to possible external domino effects [40, 41]. In own interdependent decisions, decision-makers may only want to maximize their benefits. In any case, it is very difficult to simultaneously maximize the benefits of all the decision-makers due to lack of information and myopic benefit conflicts. Nonetheless, protection decisions involving multiple decision-makers may be solved by using game-theoretical approaches.

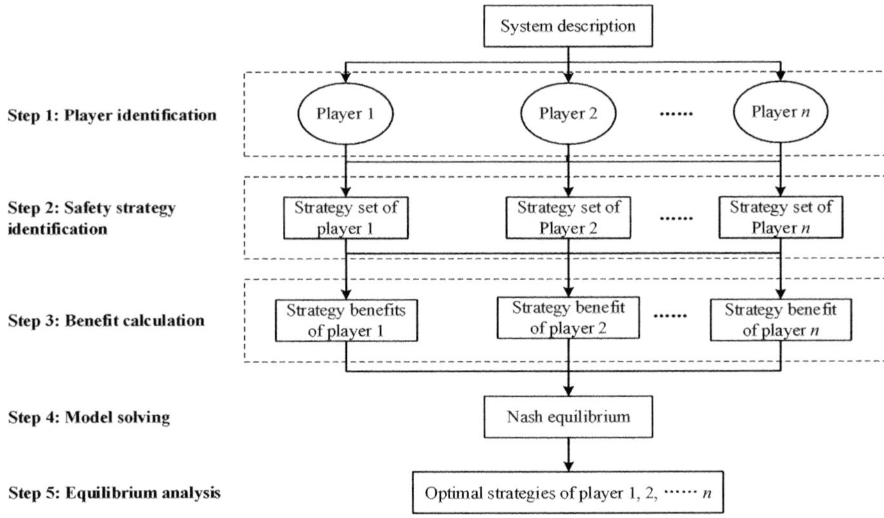

Fig. 6.8 An illustrative flowchart of a theoretical game approach (Chen et al. [1])

Game theory consists of mathematical models of strategic interaction among different decision-makers. In a game-theoretical model, decision-makers are always assumed to be intelligent and rational players who aim to maximize their benefits [42]. By solving the game, an equilibrium (trade-off) may be obtained to balance each player's benefits. Therefore, game theory may be an ideal tool for decision-making involving multiple stakeholders. Figure 6.8 shows the illustrative flowchart of a cooperative game used to achieve an optimal protection strategy.

To apply game theory in the safety and security domain, we first need to identify the stakeholders. For instance, in a chemical cluster, the players may be the chemical companies [43]. The next step is to identify the protection strategies of all players. Each player may have a set of protection strategies. Step 3 calculates the (net) benefits (objectives) of each player. In this case, possible external domino effects and safety strategies in nearby plants can be considered as well. The benefits of the players depend on all the players' motivations and strategies. In the chemical cluster case, the objective of each plant may be minimizing the total costs of damages or maximizing the net benefits. Based on the calculation of potential losses or benefits, a Nash equilibrium in which each player adopts its strictly dominant strategy may be obtained. If the equilibrium is not a pure strategy, Pareto optimal strategy of each player can be obtained.

6.3 A Cost–Benefit Analysis of Domino Effect Management

Section 6.2 introduces six economic approaches that may be used for decision-making on safety and security strategies. To deal with possible domino effects, a cost–benefit approach can be developed for decision-making on protection strategies.

6.3.1 Protection Strategy Cost

A protection strategy consists of one or more safety and security measures. To implement a protection strategy or update existing protection systems, a cost analysis is indispensable since the investment in protection strategies should meet a budget. In a protection cost analysis, all the costs related to choosing, developing, and implementing a protection strategy should be considered. As shown in Fig. 6.2, some protection costs consist of investments that only occur at the initial stage, such as initial costs and installation costs, and other costs that reoccur throughout the operation time of the measures [44]. The sub-categories of protection costs are described in Table 6.1.

As already mentioned, the initial cost does not need to be discounted to present values. In contrast, the costs that reoccur throughout the life of the protection measures (e.g., operation, maintenance, and inspection) should be discounted to present values [31]. As a result, the present value of costs ($PVC_{i,j}$) caused by the implementation of the j-th safety or security measure in a protection strategy i is the sum of the initiation costs, installation cost, and the discounted present value of reoccurring costs, as follows:

$$PVC_{i,j} = C_{i,j,\text{ini}} + C_{i,j,\text{ins}} + \frac{(1+r)^y - 1}{r_d(1+r_d)^y} C_{i,j,\text{reo}} \tag{6.1}$$

Table 6.1 Categories of protection strategy costs (Chen et al. [1])

Cost category	Subcategories
Initiation	Investigation, selection and design material, training, changing guidelines, and informing
Installation	Production loss, start-up, equipment, installation team
Operation	Utility consumption and labor
Maintenance	Material, maintenance team, production loss, start-up
Inspection	Inspection team
Logistics and transport	Transport and loading/unloading of hazardous materials, storage of hazardous materials, drafting control lists, protection documents
Contractor	Contractor selection, training
Other	Office furniture, insurance

$$C_{i,j,\text{reo}} = C_{i,j,\text{ope}} + C_{i,j,\text{mai}} + C_{i,j,\text{ins}} + C_{i,j,\text{log}} + C_{i,j,\text{con}} + C_{i,j,\text{oth}} \quad (6.2)$$

where $C_{i,j,\text{ini}}$ denotes the initial costs of measure j in strategy i, $C_{i,j,\text{ins}}$ represents the installation costs of measure j in strategy i, $C_{i,j,\text{reo}}$ concerns the installation costs of measure j in strategy i, $C_{i,j,ope}$ represents the annual operation costs of measure j in strategy i, $C_{i,j,\text{mai}}$ denotes the annual maintenance costs of measure j in strategy i, $C_{i,j,\text{ins}}$ represents the annual inspection costs of measure j in strategy i, $C_{i,j,log}$ denotes the annual logistics and transport costs of measure j in strategy i, $C_{i,j,\text{con}}$ represents the annual contractor costs of measure j in strategy i, $C_{i,j,\text{oth}}$ represents the annual other costs of measure j in strategy i, r_d is the discount rate, y is the number of years that the protection measures can operate. Consequently, the cost of a protection strategy should be the sum of the cost of each protection measure, as follows:

$$PVC_i = \sum_{j=1}^{J} PVC_{i,j} \quad (6.3)$$

PVC_i represents the present value of the cost of protection strategy i, J is the total number of protection measures in protection strategy i.

6.3.2 *The Costs of Domino Effects*

The costs of domino effects refer to the potential financial losses caused by domino effects. The losses of domino effects should consider both the direct losses caused by the initiating accident and the losses induced by the subsequent domino effect. Possible intentional attacks and accidental scenarios that may induce domino effects should be considered. There may be multiple attack scenarios in terms of an adversary since an intelligent and strategic adversary may adapt to changing circumstances in terms of protection measures. Considering K attack scenarios and U accidental scenarios are identified in a chemical plant, the overall expected losses caused by the k-th ($k = 1, 2, 3, \ldots, K$) attack scenario and the u-th ($u = 1, 2, 3, \ldots, U$) accidental scenario can be simplified as the sum product of the installations' damage probabilities and their losses:

$$L_k = \sum_{n=1}^{N} P_{k,n} \cdot L_{k,n} \quad (6.4)$$

$$L_u = \sum_{n=1}^{N} P_{u,n} \cdot L_{u,n} \quad (6.5)$$

L_k represents the total loss caused by attack k. $P_{k,n}$ denotes the damage probability of installation n in attack scenario k, $L_{k,n}$ represents the loss of installation n in attack k; L_u represents the total loss caused by accidental scenario u. $P_{u,n}$ denotes the damage probability of installation n in accident scenario u, $L_{u,n}$ represents the loss of installation n in accident u.

The losses caused by domino effects ideally should consider fatalities, injuries, property loss, and any other influences such as psychological and political effects [45]. The casualties include not only the direct losses that are immediately visible and tangible but also the indirect losses that are intangible and invisible [15, 46]. The direct avoided losses refer to the losses caused by damage to installations, products, equipment, medical expenses, paying fines, and insurance premium rise. The indirect losses consist of capacity losses, production schemes, recruitment, and wage costs, for instance [28]. The estimation of indirect losses is more difficult than that of direct losses because they are much more hidden or invisible. If these hidden or invisible losses are ignored, the total loss caused by domino effects may be largely underestimated [46]. Nonetheless, several methods for the estimation of indirect losses are available in the literature. A straightforward method is to use the ratio of indirect to direct loss. In that case, the indirect losses can be expressed as the product of direct losses and the loss ratio. However, the ratio varies in academic literature, leading to difficulty for selecting a suitable ratio. For instance, a commonly used loss ratio of 4 was recommended based on an analysis of 7500 accidents, while a range of 1–20 was also proposed for different industrial sectors [25]. Besides, Reniers and Brijs [44] proposed eight categories of losses caused by major accidents. To address the losses related to intentional attacks, reputation costs, symbolism-, psychological-, and political financial effects may also be considered [15]. According to past studies, losses caused by domino effects can be divided into eleven categories, as shown in Table 6.2.

Table 6.2 Categories of protection costs (Reniers and Van Erp [15])

Cost category	Subcategories
Supply chain	Production, start-up, schedule
Damage	Damage to own material/property, other companies' material/property, surrounding living areas, public material/property
Legal	Fines, interim lawyers, specialized lawyers, internal research team, experts at hearings, legislation, permit, and license
Insurance	Insurance premium
Human	Compensation victims, injured employees, recruitment,
Environmental	Environmental damage and clean-up
Personnel	Productivity of personnel, training of new or temporary employees, wages
Medical	Medical treatment at the location, medical treatment in hospitals and revalidation, using medical equipment and devices, medical transport
Intervention	The service from the fire department, police department, or ambulance
Investigation	Accident investigation
Other	Reputational, symbolic, psychological, and political effects

Therefore, the loss of an installation n caused by intentional attack k ($L_{k,n}$) or accident scenario u ($L_{u,n}$) can be estimated [2], as follows:

$$L_{k,n} = L_{k,n,sup} + L_{k,n,dam} + L_{k,n,leg} + L_{k,n,ins} + L_{k,n,hum} + L_{k,n,env} + L_{k,n,per} + L_{k,n,med} + L_{k,n,int} + L_{k,n,inv} + L_{k,n,oth} \tag{6.6}$$

$$L_{u,n} = L_{u,n,sup} + L_{u,n,dam} + L_{u,n,leg} + L_{u,n,ins} + L_{u,n,hum} + L_{u,n,env} + L_{u,n,per} + L_{u,n,med} + L_{u,n,int} + L_{u,n,inv} + L_{u,n,oth} \tag{6.7}$$

where $L_{k,n,sup}$ represents the supply chain loss of installation n caused by attack k, $L_{k,n,dam}$ denotes the damage loss of installation n caused by attack k, $L_{k,n,leg}$ represents the legal loss of installation n caused by attack k, $L_{k,n,ins}$ denotes the insurance loss of installation n caused by attack k, $L_{k,n,hum}$ denotes the human loss of installation n caused by attack k, $L_{k,n,env}$ represents the environmental loss of installation n caused by attack k, $L_{k,n,per}$ represents the personnel loss of installation n caused by attack k, $L_{k,n,med}$ represents the medical loss of installation n caused by attack k, $L_{k,n,int}$ denotes the intervention loss of installation n caused by attack k, $L_{k,n,rep}$ denotes the reputation loss of installation n caused by attack k, $L_{k,n,inv}$ denotes the accident investigation and the cleanup loss of installation n caused by attack k, $L_{k,n,oth}$ represents other losses of installation n caused by attack k. $L_{u,n,sup}$ represents the supply chain loss of installation n caused by accident scenario u, $L_{u,n,dam}$ denotes the damage loss of installation n caused by accident scenario u, $L_{u,n,leg}$ represents the legal loss of installation n caused by accident scenario u, $L_{u,n,ins}$ denotes the insurance loss of installation n caused by accident scenario u, $L_{u,n,hum}$ denotes the human loss of installation n caused by accident scenario u, $L_{u,n,env}$ represents the environmental loss of installation n caused by accident scenario u, $L_{u,n,per}$ represents the personnel loss of installation n caused by accident scenario u, $L_{u,n,med}$ represents the medical loss of installation n caused by accident scenario u, $L_{u,n,int}$ denotes the intervention loss of installation n caused by accident scenario u, $L_{u,n,rep}$ denotes the reputation loss of installation n caused by accident scenario u, $L_{u,n,inv}$ denotes the accident investigation and the cleanup loss of installation n caused by accident scenario u, $L_{u,n,oth}$ represents other losses of installation n caused by accident scenario u.

6.3.3 Net Benefits Analysis

According to the hypothetical benefit assumption illustrated in Sect. 6.2.3, the benefits of an integrated protection strategy can be expressed as the difference between losses of domino accidents without and with the implementation of the protection strategy. To calculate the benefits of a protection strategy, a baseline ($k = 0$) can be defined as the strategy without any safety or security measures or the initial strategy before protection upgrade. Therefore, the benefits of a protection strategy i for a particular scenario can be defined, as follows:

$$B_{i,k} = L_{0,k} - L_{i,k} \tag{6.8}$$

$$B_{i,u} = L_{0,u} - L_{i,u} \tag{6.9}$$

$B_{i,k}$ represents the benefit of protection strategy i for a special attack scenario k, $L_{0,k}$ denotes the expected loss caused by attack scenario k under the protection of baseline strategy 0, $L_{i,k}$ represents the expected loss caused by attack scenario k under the protection of strategy i; $B_{i,u}$ denotes the benefit of protection strategy i for a particular accidental scenario u, $L_{0,u}$ represents the expected loss caused by accidental scenario u under the protection of baseline strategy 0, $L_{i,u}$ denotes the expected loss caused by accidental scenario u under the protection of strategy i.

In this chapter, possible intentional domino effects are considered. The attacker is assumed as a benefit maximizer aiming to maximize the damage while the defender aims to minimize the losses caused by both intentional and unintentional domino effects. According to the Stackelberg leadership model [47, 48], the defender can be considered the 'leader' who takes the prior decision on protection strategy while the adversary is regarded as the 'follower' who knows the complete protection information before launching an attack. For a protection strategy i, the adversary would adapt to the protection by selecting an attack scenario k maximizing the loss $L_{i,k}$. In other words, the benefit of a protection strategy i for intentional domino effects should be represented by the attack scenario with the minimum benefit. In that case, the total benefit of protection i for intentional and unintentional domino effects can be obtained, as follows:

$$B_i = \min_k B_{i,k} + \sum_{u=1}^{U} B_{i,u} \tag{6.10}$$

B_i represents the expected benefit of protection strategy i. Furtherly, the present value of benefits (PVB_i) of protection strategy i can be obtained, as follows:

$$PVB_i = \frac{(1+r_d)^y - 1}{r_d(1+r_d)^y} B_i \tag{6.11}$$

Consequently, the so-called proportion factor (PF) representing the ratio of the costs to the benefits can also be obtained [49], as follows:

$$PF_i = \frac{PVC_i}{PVB_i} \tag{6.12}$$

PF_i is compared with the disproportion factor (DF) to determine whether the protection strategy i is "grossly disproportionate" or not. A protection strategy i is usually recommended if PF_i is less than DF. The value of the DF is always no less than 3 and not more than 30 [49]. For more information on DF, the reader can be referred to Talarico and Reniers [49] and Reniers and Van Erp [15].

6.4 Optimization Algorithm

Based on the net present value of benefits (NPVB) obtained in cost–benefit analysis, we can judge whether a protection strategy is profitable. However, companies usually face budget limitations, and there may be thousands of strategies with an NPVB value greater than zero. Therefore, we need to find out the most profitable strategies from all the possible strategies under budget limitations.

The decision-making on the investment and allocation of protection measures in chemical industrial areas can be simplified, according to Reniers and Sorensen [50], as the well-known "Knapsack problem". As a result, a non-linear optimization model can be obtained as follows:

$$\begin{cases} \max\limits_{n} NPVB_i \\ C_i \leq C_{\text{Budget}} \\ i = 1, 2, \ldots, I \end{cases} \tag{6.13}$$

The formula indicates that the protection strategy with the maximum *NPVB* value should be the optimal strategy, and the implementation cost of the protection strategy should not exceed the protection budget (C_{Budget}). To find out the optimization strategy, an optimization algorithm based on the "maximin" strategy called "PROTOPT" for PROTection OPTimization, is proposed to sequentially allocate protection measures, as shown in Fig. 6.9.

As shown in Fig. 6.9, the PROTOPT algorithm mainly consists of three steps: an implementation cost analysis, a benefit analysis, and an evaluation of *NPVB*. Based on all the available protection measures, we need to calculate the implementation cost of each one. The protection measure with an implementation cost lower than the budget will be the input of the second step. In the second step, the performance of each selected protection measure should be assessed based on vulnerability assessment models provided in Chaps. 2–5. Then a benefit analysis will be conducted to obtain the *NPVB* value of the protection measure. In the last step, the protection measures with the maximum *PVB* should be selected as one of the measures of the optimal strategy. The above calculation loop will continue until $PF > DF$ or the total implementation cost is greater than the budget [49]. Applying the optimization algorithm, we can obtain the optimal protection strategy under a protection budget and obtain a recommended protection cost based on *DF*.

6.5 Conclusions

Following the integrated management framework developed in Chap. 5, this chapter introduces an economic approach to support decision-making on investment and allocation of protection measures. In this chapter, some well-known economic approach methods used in the safety and security domain are presented in a nutshell, including

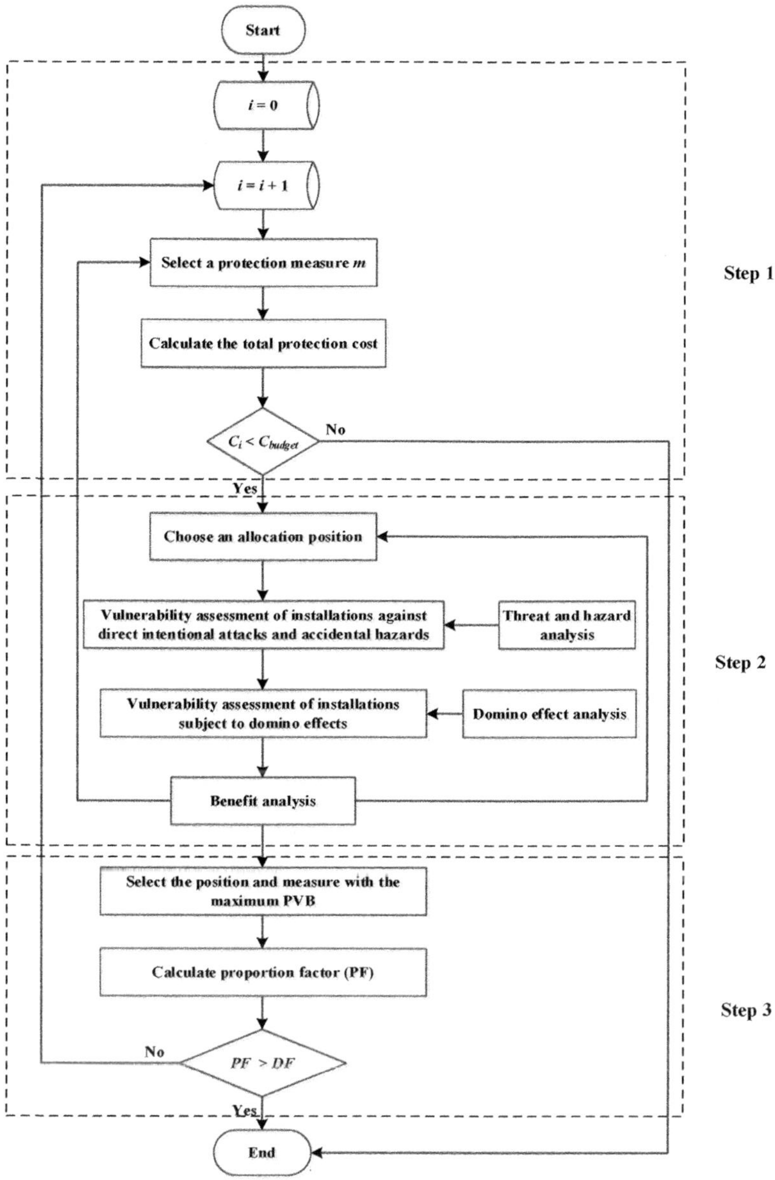

Fig. 6.9 The "PROTOPT" algorithm to achieve an optimal protection strategy (Talarico and Reniers [49] and Chen et al. [2])

risk-based optimization, minimal total safety cost approach, cost–benefit analysis, cost-effectiveness analysis, multi-objective optimization, and the game-theoretical approach. Based on a cost–benefit analysis and disproportion factor, a decision-making approach for preventing and mitigating intentional and unintentional domino effects is developed. Based on the approach, a sequential allocation algorithm is developed to obtain the optimal protection strategy under the limitation of the protection budget. If no protection budget is provided, an optimal protection budget can also be achieved by using the developed approach.

References

1. Chen C, Reniers G, Khakzad N, Yang M (2021) Operational safety economics: foundations, current approaches and paths for future research. Saf Sci 141
2. Chen C, Reniers G, Khakzad N (2020) Cost-benefit management of intentional domino effects in chemical industrial areas. Process Saf Environ Prot 134:392–405
3. Buncefield Major Incident Investigation Board (2008) The Buncefield incident, 11 December 2005: the final report of the Major Incident Investigation Board. Health and Safety Executive
4. Johnson D (2013) Vapour cloud explosion at the IOC terminal in Jaipur. Loss Prev Bull (229)
5. Reniers G, Soudan K (2010) A game-theoretical approach for reciprocal security-related prevention investment decisions. Reliab Eng Syst Saf 95(1):1–9
6. Reniers GLL, Audenaert A (2014) Preparing for major terrorist attacks against chemical clusters: intelligently planning protection measures w.r.t. domino effects. Process Saf Environ Prot 92(6):583–589
7. Janssens J, Talarico L, Reniers G, Sörensen K (2015) A decision model to allocate protective safety barriers and mitigate domino effects. Reliab Eng Syst Saf 143:44–52
8. Landucci G, Reniers G, Cozzani V, Salzano E (2015) Vulnerability of industrial facilities to attacks with improvised explosive devices aimed attriggering domino scenarios. Reliab Eng Syst Saf 143:53–62
9. Zhou J, Reniers G (2016) Petri-net based simulation analysis for emergency response to multiple simultaneous large-scale fires. J Loss Prev Process Ind 40:554–562
10. Hosseinnia B, Khakzad N, Reniers G (2018) An emergency response decision matrix against terrorist attacks with improvised device in chemical clusters. Int J Saf Secur Eng 8(2):187–199
11. Khakzad N, Reniers G (2019) Low-capacity utilization of process plants: a cost-robust approach to tackle man-made domino effects. Reliab Eng Syst Saf 191
12. Birk AM (2014) Cost-effective application of thermal protection on LPG road transport tanks for risk reduction due to hot BLEVE incidents. Risk Anal 34(6):1139–1148
13. Paltrinieri N, Bonvicini S, Spadoni G, Cozzani V (2012) Cost-benefit analysis of passive fire protections in road LPG transportation. Risk Anal 32(2):200–219; discussion 220–203
14. Poole RW (2008) Toward risk-based aviation security policy. OECD/ITF Joint Transport Research Centre Discussion Paper
15. Reniers GL, Van Erp HN (2016) Operational safety economics: a practical approach focused on the chemical and process industries. Wiley
16. Spengler J (1968) The economics of safety. Law Contemp Probs 33:619
17. Oi WY (1974) On the economics of industrial safety. Law Contemp Probl 38(4):669–699
18. Chen C, Reniers G (2020) Chemical industry in China: the current status, safety problems, and pathways for future sustainable development. Saf Sci 128
19. Kröger W, Fischer P (2000) Balancing safety and economics. Nucl Eng Des 195(1):101–108
20. Annas GJ (1983) Nuclear power: safety and economics. Am J Public Health 73(9):1099–1100
21. Chen C, Reniers G (2018) Risk assessment of processes and products in industrial biotechnology. Adv Biochem Eng Biotechnol

22. Taylor JR (2003) Risk analysis for process plant, pipelines and transport. Routledge
23. Cozzani V, Gubinelli G, Antonioni G, Spadoni G, Zanelli S (2005) The assessment of risk caused by domino effect in quantitative area risk analysis. J Hazard Mater 127(1–3):14–30
24. Jonkman SN, van Gelder PHAJM, Vrijling JK (2003) An overview of quantitative risk measures for loss of life and economic damage. J Hazard Mater 99(1):1–30
25. Dorman P (2000) The economics of safety, health, and well-being at work: an overview. ILO Geneva
26. Baladeh, A. E., Cheraghi, M., & Khakzad, N. (2019). A multi-objective model to optimal selection of safety measures in oil and gas facilities. Process Safety and Environmental Protection, 125, 71–82.
27. Sun L, Paez O, Lee D, Salem S, Daraiseh N (2006) Estimating the uninsured costs of work-related accidents, Part I: A systematic review. Theor Iss Ergon Sci 7(3):227–245
28. Gavious A, Mizrahi S, Shani Y, Minchuk Y (2009) The costs of industrial accidents for the organization: developing methods and tools for evaluation and cost–benefit analysis of investment in safety. J Loss Prev Process Ind 22(4):434–438
29. Arends BJ, Jonkman SN, Vrijling JK, van Gelder PHAJM (2005) Evaluation of tunnel safety: towards an economic safety optimum. Reliab Eng Syst Saf 90(2–3):217–228
30. Aven T (2015) Risk analysis. Wiley
31. Chen C, Reniers G (2020) Economic approaches for making prevention and safety investment decisions in the process industry
32. HSE principles for Cost Benefit Analysis (CBA) in support of ALARP decisions. https://www.hse.gov.uk/managing/theory/alarpcba.htm#:~:text=CBA%20aids%20the%20decision%20making,a%20comparison%20of%20like%20quantities.&text=For%20many%20ALARP%20decisions%2C%20the,costs%20and%20benefits%20may%20suffice. Accessed 12 March 2021
33. Levin HM (1995) Cost-effectiveness analysis. Int Encycl Econ Educ 2:381–386
34. Murray CJ, Evans DB, Acharya A, Baltussen RMJHe (2000) Development of WHO guidelines on generalized cost-effectiveness analysis. 9(3):235–251
35. Briggs A, Fenn P (1997) Trying to do better than average: a commentary on 'statistical inference for cost-effectiveness ratios.' Health Econ 6(5):491–495
36. Miettinen K (2012) Nonlinear multiobjective optimization, vol 12. Springer Science & Business Media
37. Eini S, Shahhosseini HR, Javidi M, Sharifzadeh M, Rashtchian D (2016) Inherently safe and economically optimal design using multi objective optimization: the case of a refrigeration cycle. Process Saf Environ Prot 104:254–267
38. Ramadhan NJ, Wan YK, Ng RTL, Ng DKS, Hassim MH, Aviso KB, Tan RR (2014) Life cycle optimisation (LCO) of product systems with consideration of occupational fatalities. Process Saf Environ Prot 92(5):390–405
39. Gao Y, Fan Y, Wang J (2020) Assessing the safety regulatory process of compliance-based paradigm in China using a signalling game model. Saf Sci 126
40. Reniers GLL, Dullaert W, Ale BJM, Soudan K (2005) Developing an external domino accident prevention framework: Hazwim. J Loss Prev Process Ind 18(3):127–138
41. Hosseinnia B, Khakzad N, Reniers G (2018) Multi-plant emergency response for tackling major accidents in chemical industrial areas. Saf Sci 102:275–289
42. Myerson RB (2013) Game theory. Harvard University Press
43. Reniers G, Pavlova Y (2013) Using game theory to improve safety within chemical industrial parks. Springer
44. Reniers G, Brijs T (2014) Major accident management in the process industry: an expert tool called CESMA for intelligent allocation of prevention investments. Process Saf Environ Prot 92(6):779–788
45. Stewart MG, Mueller J (2011) Cost-benefit analysis of advanced imaging technology full body scanners for airline passenger security screening. J Homel Secur Emerg Manag 8(1)
46. Jallon R, Imbeau D, de Marcellis-Warin N (2011) Development of an indirect-cost calculation model suitable for workplace use. J Saf Res 42(3):149–164

47. Kroshl WM, Sarkani S, Mazzuchi TA (2015) Efficient allocation of resources for defense of spatially distributed networks using agent-based simulation. Risk Anal 35(9):1690–1705
48. Pita J, Jain M, Ordóñez F, Tambe M, Kraus S, Magori-Cohen R (2009) Effective solutions for real-world stackelberg games: when agents must deal with human uncertainties. In: Proceedings of the 8th international conference on autonomous agents and multiagent systems-Volume 1. International Foundation for Autonomous Agents and Multiagent Systems, pp 369–376
49. Talarico L, Reniers G (2016) Risk-informed decision making of safety investments by using the disproportion factor. Process Saf Environ Prot 100:117–130
50. Reniers GL, Sorensen K (2013) An approach for optimal allocation of safety resources: using the knapsack problem to take aggregated cost-efficient preventive measures. Risk Anal 33(11):2056–2067

Chapter 7
A Resilience-Based Approach for the Prevention and Mitigation of Domino Effects

7.1 Introduction

The chemical industry is pivotal for welfare, industrial production and daily life by providing various chemical products such as gasoline, benzene, and natural gas.[1] Most of these chemical products are usually flammable, combustible, and toxic, and are stored, transferred, and processed via different equipment and installations within a chemical plant. These hazardous materials make process plants vulnerable to disruptions, with the possibility of major accidents such as fire, explosion, and/or hazardous release [3–8].

Several types of disruptions are possible: unintentional accidents, natural disasters, and intentional attacks [3, 9, 10]. Unplanned disruptions are induced by unintentional events such as mechanical failure, corrosion, fatigue, and human errors. For instance, the Intercontinental Terminal Company Tank Fire in March 2019 at Deer Park in the US was triggered by an unloading operation [11]. Compared with accidental disruptions, natural hazard-related disasters may lead to more severe consequences due to the damage of multiple chemical facilities at once, safety barriers, and other emergency response infrastructures. The damage to industrial facilities induced by natural disasters is called the "Natech event" [12, 13]. For example, Hurricane Harvey in 2017 led to the release of the contents of at least 18 hazardous storage tanks in Texas [14, 15]. Both accidental disruptions and the disruptions caused by natural hazards are unintentional, while intentional attacks may aim to cause damage to chemical facilities deliberately. For instance, on June 26, 2015, two tanks in a France chemical plant were damaged due to an intentional attack by using explosive devices [16].

Many studies dealing with unintentional and intentional disruptions have been conducted. Types of various measures were explored w.r.t. disruptions such as

[1] This chapter is mainly based on two publications: Chen et al. [1, 2].

C. Chen et al., *Integrating Safety and Security Management to Protect Chemical Industrial Areas from Domino Effects*, Springer Series in Reliability Engineering,
https://doi.org/10.1007/978-3-030-88911-1_7

inherent safety [17–21], hazard identification [22, 23], safety risk assessment [24–27], Natech risk assessment [10, 28, 29], security risk assessment [30–33], safety barrier management [34–36], security measure management [37–39], emergency response [40–42], integrated safety and security management [43–45], etc. No matter what the disruption is, the primary undesired scenario caused by the disruption may escalate to installations nearby, leading to a chain of accidents, resulting in the overall consequences more severe than the primary event, the so-called "domino effects" [46]. Chapters 2, 3, 4 and 5 provide models for assessing domino effects, and Chaps. 5, 6 develop an approach to manage escalation effects using safety and security measures. Safety and security management aims to reduce the likelihood and consequences caused by undesired events for avoiding and mitigating human loss, economic loss, environmental loss, etc. However, domino effects caused by intentional attacks or natural disasters may be unpredictable and unpreventable due to the simultaneous failure of multiple installations. Only considering safety and security measures may not be sufficient to deal with unpredictable or unpreventable domino effects. This chapter thus provides a resilience approach to manage domino effects, considering safety measures, security measures as well as adaptation measures, and restoration measures. In this chapter, a methodology for quantifying the resilience of chemical plants, considering the dynamic stochastic evolution of disruptions due to domino effects, adaptation performance, and the dynamic restoration process. This chapter is organized as follows: Sect. 7.2 introduces the concept of resilience and defines chemical plant resilience, and introduces the possible measures to enhance the resilience of chemical plants. A stochastic dynamic methodology for quantifying the chemical plant resilience is elaborated in Sect. 7.3. Section 7.4 develops an algorithm to obtain chemical resilience. Finally, the conclusions drawn from this study are present in Sect. 7.5.

7.2 Chemical Plant Resilience

7.2.1 Resilience Concept

The term "resilience" originated from the Latin word "resiliere" which refers to the bounce back from disruptions [47]. Based on the original meaning of "resiliere", a narrow definition of resilience can be obtained: "the ability of a system to recover to a normal condition or a new stable equilibrium after an event disrupts from its original state" [48]. Unlike the concept focusing on the recovery capability of a system, Holling [49] defined resilience as "the ability to absorb disruptions without dramatic change. The definition thus focuses more on the ability of a system to withstand disruptions and maintain its functions". Integrating the elements in the above two definitions, National Infrastructure Advisory Council [50] defined a more comprehensive definition: the ability to predict, absorb, adapt, and/or quickly recover from disruptions. Many other definitions of resilience are also available in the literature

[6, 47, 48], but no identical definition of resilience exists in the academic domain [47]. According to these definitions, the abilities (metrics) of a resilient system for responding to unexpected disruptions can be summarized as follows [51–53]:

- Absorptive ability: the ability of a system to resist, absorb, or withstand the impact of disruptive events;
- Adaptive ability: the ability of a system to adapt itself to maintain its operational performance after disruptions without any recovery activity;
- Recovery ability: the capability of a system to repair or restore damages from a disruption to recover the loss performance of the system, making the system reach a new stable equilibrium.

In recent years, resilience is becoming a more active and substantial research topic in the safety and security domain. Resilience engineering aims to enhance a system's capabilities to absorb, adapt, and recover from a disruption, and reduce the impacts of the disruptions on the system's performance. Safety and security management may be used to enhance absorption capability while has no direct impacts on adaption and recovery capabilities. Safety and security management is not as wide as that of resilience management/engineering. In terms of unpredictable or indefensible threats (e.g., terrorist attacks and natural disasters), enhancing the resilience capability is an ideal approach to reduce the losses caused by disruptions and to quickly recover a system's performance. [54].

The advancements of resilience engineering can considerably contribute to process safety and security [7, 53, 55–57]. Past research attempts on resilience in the process industry identified process resilience influence factors [57, 58], resilience hazards [59–61], and hired quantification of process resilience using Bayesian network [6, 54, 62]. However, little attention has been paid to chemical plant resilience in which escalation effects may play an essential role [53, 63]. Consequently, a new definition of chemical plant resilience will be provided in this chapter.

7.2.2 A Definition of Chemical Plant Resilience

Resilience has been used in various industries and systems, such as ecological systems [49], communities [64], and container terminals [65]. In light of possible catastrophic effects in the chemical and process industry, we define chemical plant resilience as the ability of a chemical plant to resist, mitigate, adapt and recover from undesired events to maintain its operation. Unlike safety and security in the process and chemical industry that aims to prevent and mitigate undesired events, chemical plant resilience intends to enhance a system's capability to anticipate and prepare for disruption and its ability to adapt and recover from the disruption. In other words, chemical plant resilience aims to prevent disruptions, withstand and mitigate their consequences during the evolution of disruptions and quickly recover their functions after disruptions. To enhance chemical plant resilience, Resilience measures may be

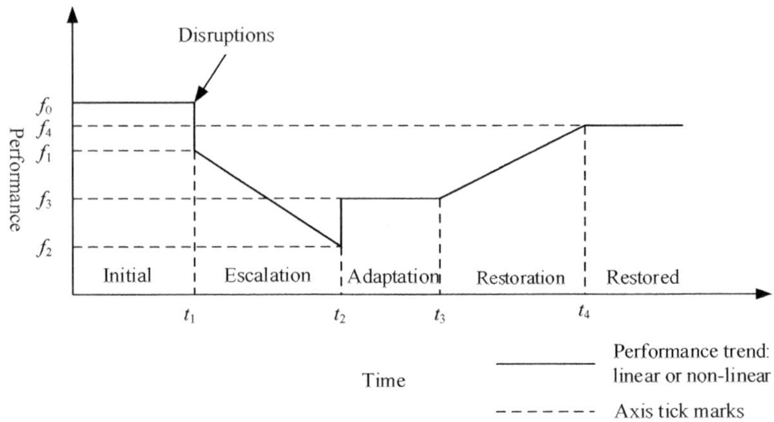

Fig. 7.1 Chemical plant performance varies over time (adapted from Henry and Emmanuel Ramirez-Marquez [66])

implemented in different stages to resist the impacts of an undesired event, mitigate the consequences by preventing possible domino effects, and adjust operation strategies to improve the operation performance before recovery and to rapidly recover the plants. Figure 7.1 shows the chemical plant performance response to a disruption.

As shown in Fig. 7.1, the chemical plant performance evolves, and the entire process from a disruption to restoration is called a resilience evolution scenario. For a disruption event, there may be multiple resilience evolution scenarios since the uncertainties in the vulnerability of installations, hazardous scenario escalation, emergency response, adaptation strategies, and restoration strategies. A resilience evolution scenario of a chemical plant can be divided into six stages: initial stage, disruption stage, escalation stage, adaptation stage, restoration stage, and restored stage. The chemical plant performance decreases in the disruption stage and the escalation stage while increasing in the adaptation and restoration stage. Before the occurrence of a disruption event, the chemical plant is in the initial stage in which the performance is the initial value f_0. When a disruption occurs, the chemical plant performance may decrease immediately and cause major accident scenarios due to the damage to one or more hazardous facilities. Primary major accident scenarios may trigger secondary and higher-order major accident scenarios, the so-called domino effects, damaging the installations nearby and further decreasing the performance of the chemical plant. When the escalation of domino effects ends at t_2, the residual performance reduces to the minimum value f_2. Then, the chemical plant may change its operation strategies to adapt to the new state and thus improve its performance. For instance, a chemical plant may utilize reserve equipment or adjust production processes to partly or fully offset the reduced performance. The last resilience stage is restoration, in which the performance of the chemical plant is recovered by repairing or rebuilding the damaged installations ($t_3 \sim t_4$). In Fig. 7.1 the performance is not fully recovered after the restoration stage at t_4. In general, the performance at the end of the restoration stage may be in three possible states: the recovered performance is equal to the initial performance (fully recovered), the recovered performance is less than the initial performance (partly recovered, usually, this is the case in industrial practice, see also Fig. 7.1), and the recovered performance is greater than the initial performance (overly recovered). In real cases, the performance of a recovered chemical plant may then be different from the plant in the initial stage.

7.2.3 *Resilience Metrics*

According to the definition of chemical plant resilience, resilience metrics should characterize the ability to withstand and resist disruptions, the ability to prevent the escalation of a disruption event, the ability to adapt to the disrupted condition, and the ability to quickly recover from the disruption. Combining the definition and the performance curve in Fig. 7.1, it can be demonstrated that the resilience of chemical plants mainly depends on two factors: chemical plant performance and evolution time. As a result, the essential step to quantify chemical plant resilience is to quantify the change of chemical plant performance over time during a resilience evolution scenario. Based on the resilience framework proposed by Bruneau et al. [64], the expected resilience loss of a chemical plant can be expressed as the expected degradation in performance over time, and resilience thus can be represented by the expected resilience loss divided by the planned resilience. Therefore, we define the chemical plant resilience as a dimensionless ratio, as follows:

$$R = \frac{\int_{t_1}^{t_4} f(t)}{f(t_0)(t_4 - t_1)} \tag{7.1}$$

$f(t)$ is the chemical plant performance function which depends on resilience evolution time (t); $f(t_0)$ is the performance at the initial stage (planned performance). The numerator of Eq. (7.1) represents the accumulation of chemical plant performance $f(t)$ during a resilience evolution process (between t_1 (disruption) and t_4 (fully recovered)). The denominator represents the accumulation of the planned performance $f(t_0)$ from t_1 to t_4.

Equation (7.1) only addresses one resilience evolution scenario. Due to the uncertainties in resilience evolution, there may be many resilience evolution scenarios that can be seen as different performance curves. Considering a total of X resilience evolution scenarios and the maximum value of t_4 of all the scenarios being t_{max}, the resilience metrics considering uncertainties can be obtained, as follows:

$$R = \frac{1}{X} \sum_{i=1}^{X} \frac{\int_{t_1}^{t_{max}} f(t)}{f(t_0)(t_{max} - t_1)} \tag{7.2}$$

In Eq. 7.2, t_4 is substituted with t_{max} to unify the integration interval and avoid overestimating the resilience evolution scenarios with longer resilience evolution time or underestimating the resilience evolution scenarios with shorter resilience evolution time. According to Eq. 7.2, the most resilient chemical plant (R is equal to 1) is an ideal condition in which a disruption event does not lead to any performance degradation. In such case, the impact of the disruption on the chemical plant is fully absorbed, and the chemical plant continuously maintains its initial performance. On the contrary, R is equal to zero when the chemical plant is damaged by the disruption, the performance immediately drops to zero, and adaptation and recovery are impossible. Usually, the resilience value of R is between 0 and 1. It should be marked that t_4 is the time the system is fully recovered if the performance at the end of the restoration exceeds the initial performance. In that case, the maximum recovered performance at t_4 can not exceed its initial performance, and R is no more than 1. Although the resilience metrics is established by the case of chemical plants, it may be applied to other infrastructure systems with catastrophic effects by substituting infrastructure system performance functions for the chemical plant performance function $f(t)$.

7.2.4 Capabilities of Chemical Plant Resilience

According to the performance evolution curve shown in Fig. 7.2 and the resilience metrics shown in Eq. (7.2), chemical plant resilience capabilities consist of resistance capability, mitigation capability, adaptation capability, and restoration capability. Therefore, the resilience of a chemical plant can be improved by resistance measures,

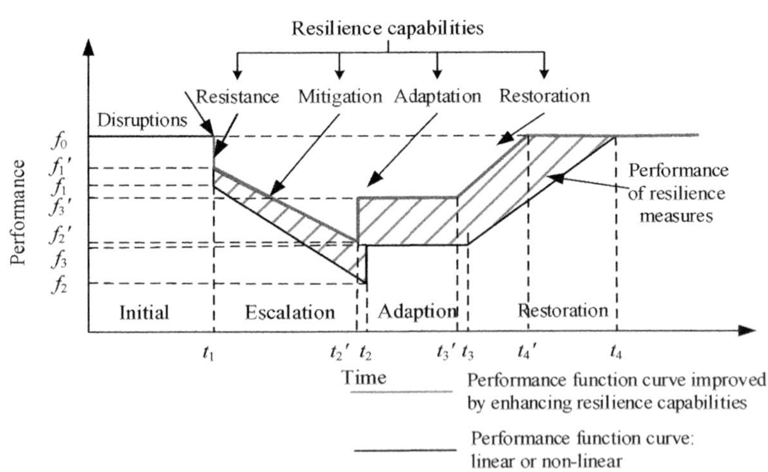

Fig. 7.2 Chemical plant performance improved by resilience capabilities (Chen et al. [1])

mitigation resilience, adaptation measures and restoration measures, as shown in Fig. 7.2.

(1) **Resilience capability**

Resistance capability is the capability to withstand disruption events, avoid failure of installations, and maintain performance. Similar to the prevention capability in safety and security management, various measures can improve resistance capability, and different measures may be taken for tackling different disruptions. For instance, installing lightning masts around installations and installing air terminals on the installations can prevent the damage of installations caused by lightning strikes [5, 67]; while detection measures such as closed-circuit television (CCTV) cameras can be used to detect illegal invasions and accidental scenarios [16, 38]. By implementing resistance measures, resistance capability can be improved and thus increase the value of f_1 in Fig. 7.2, enhancing resilience.

(2) **Mitigation capability**

Mitigation capability is the capability to prevent the escalation of possible major accident scenarios caused by disruptions. Mitigation capability is essential for chemical plants with many hazardous infrastructures due to possible domino effects. In this chapter, the mitigation capability only refers to the capability to prevent and mitigate domino effects. In contrast, the capability to mitigate the performance loss in the disruption stage is considered in resistance capability. As a result, any measures that can be used to prevent and mitigate domino effects can be applied to enhance mitigation capability, such as active measures (e.g., fire sprinkler system), passive measures (e.g., fireproof coating), procedural and emergency response (e.g., firefighting) [3, 68, 69]. Under the protection of mitigation capability, possible domino effects may be prevented or mitigated, leading to a higher value of f_2 and a shorter time between t_1 and t_2, thus reducing the performance degradation and improving the resilience.

(3) **Adaptation capability**

Adaptation capability is the capability to adapt to a new operation state to wholly or partly recover the chemical plant performance before restoration. At the end of the escalation stage, chemical plants may adjust operation strategies (e.g., utilizing reserve equipment and speeding inventory turnover ratio) to partly or fully recover the chemical plant performance. To enhance adaptation capability, a chemical plant can provide reserve equipment and prepare multiple operation strategies under the lack of installations. By using these measures, the adaptation capability can be enhanced, and the value of f_3 in Fig. 7.2 can be increased, resulting in the increase of chemical plant performance before the restoration of the chemical plant.

(4) **Restoration capability**

Restoration capability refers to the capability to quickly repair or rebuild the damaged installations to recover the chemical plant performance. The restoration capability mainly depends on the time to fully recover (*TTR*). Therefore, shortening the *TTR* (t_3–t_4) can effectively reduce the lost performance and achieve a more resilient chemical plant. To shorten the restoration time, a chemical plant may formulate restoration plans, establish a maintenance team, or reach contact with construction companies. Once a disruption occurs, these precautions can help the chemical company to save time and obtain a quick recovery.

7.3 A Quantification Framework of Chemical Plant Resilience

Chemical plants are industrial infrastructures that manufacture, process, and storage chemical materials. The performance of a chemical plant mainly depends on its operation and products. For instance, hazardous material storage plants are industrial facilities for storing hazardous chemicals such as petroleum, benzene, and other chemical products. These products are delivered to end-users, process facilities, and other storage facilities. As a result, the total storage volume of the plant or the average daily chemical flow rate at the initial stage of a hazardous material storage plant can be used to represent the chemical plant performance. According to the chemical plant performance, the performance function $f(\mathrm{t})$ can be established by quantified the capability of resistance, mitigation, adaptation, and restoration.

Based on the description of chemical plant resilience capability, the next section is to develop a framework to quantify the resilience metrics by modeling these resilience capabilities. Modeling the resilience capabilities based on the performance curve is the crucial step to quantify the resilience of a chemical plant.

7.3.1 Resistance Modeling

Resistance is the capability to withstand disruptions, avoid being damaged and retain performance [52]. From the perspective of safety and security management, resistance may be deemed the antonym of vulnerability which represents the inability of an installation to withstand disruptions and the consequent failures [70]. The vulnerability of installations is usually characterized by the failure probability of installations exposed to disruption events. Therefore, the resistance capability of an installation can be represented by the probability that an installation successfully resistant to a disruption event, as follows:

$$C_r = 1 - P_f \tag{7.3}$$

C_r represents the resistance capability of the installation exposed to a disruption event, P_f represents the failure probability of an installation exposed to the disruption. In a chemical plant, there are usually multiple installations situated nearby. A disruption event such as an explosion [16] or a natural disaster may lead to simultaneous failure of multiple installations, resulting in a sudden degradation of chemical plant performance (from f_0 to f_1). f_1 is the performance of the chemical plant with undamaged installations, as shown in Fig. 7.1. According to *Cr,* the damaged installations in the disruption stage of each resilience evolution scenario can be determined by sampling random numbers based on Monte Carlo simulation. Consequently, the total performance at t_1 (f_1) can be obtained according to the damaged performance in the disruption stage (f_{di}), as follows:

$$f_1 = f_0 - f_{di} \tag{7.4}$$

To determine the failure probability of an installation, we need to conduct a vulnerability assessment for all the installations. For a specific installation, the vulnerability of an installation depends on the type and intensity of disruptions.

7.3.2 Mitigation Modeling

Domino effects are a discriminate phenomenon within the process and chemical industry whereby many hazardous installations are usually located nearby. Domino effects may occur due to the escalation of accident scenarios (e.g., fire and explosion) caused by disruptions. The spatial–temporal evolution of domino effects can lead to consequences more severe than the primary disruption. The mitigation ability of chemical plants thus refers to preventing and mitigating the escalation of major accident scenarios. When one or more hazardous installations are damaged and leads to loss of containment of hazardous materials, major hazards such as fire and explosion can occur, resulting in the nearby installations exposed to escalation vectors such as heat radiation and overpressure. Once the escalation vectors damage the nearby installations, the major accident scenarios may propagate, resulting in a chain of accidents and decreased performance. To avoid the failure of installations caused by domino effects, safety barriers such as passive barriers, active barriers, and emergency response barriers may be implemented [68, 69]. Passive fire protection measures refer to these safety measures that do not need external activation to trigger the protection functions for controlling fire or delaying fire escalation, such as fireproof coating and pressure safety valves. These barriers are based on different mechanisms and thus have different performances for improving mitigation capability. In terms of active protection measures, external activations are needed to trigger the protection function, such as the water spray system (WSS) and relief valves. The third safety barrier is emergency response actions. Emergency response actions such as firefighting teams are essential to prevent domino effects while a time-lapse is needed for the emergency response team to arrive. The emergency response measures can be considered a socio-technical system with some uncertainties since the procedural events depend on human behavior. The performance of these safety barriers has been illustrated in Chap. 5. Applying the Domino Evolution Graph (DEG) model in this study, we can obtain the time t_2 (at the end of the escalation stage) and the failure likelihood of installations due to domino effects. Based on the failure probability, random numbers should be sampled again to determine the failure installations in one resilience evolution scenario. Then the total performance (f_2) can be obtained according to the damaged installations in the disruption and escalation stages, as follows:

$$f_2 = f_1 - f_{es} \tag{7.5}$$

f_2 denotes the chemical plant performance at the end of the escalation stage, and f_{es} represents the performance loss due to the escalation of domino effects. Enhancing the mitigation capability can raise f_2 and/or decrease t_2, thus improving chemical plant resilience.

7.3.3 Adaptation Modeling

In the chapter, adaptation capability refers to operation adjustments, which can lead to improved chemical plant performance. These operation strategies include utilizing reserve installations, speeding inventory turnover ratio, adjusting chemical production strategies, etc. The first operation adjustment needs to add reserve ones for vulnerable installations. Once an installation is damaged, the reserve installation can quickly replace the damaged installations and thus increase the chemical plant performance. Speeding the inventory turnover ratio can increase the daily chemical flow rate and thus reduce the effect of damaged installations on the chemical plant performance. Adjusting chemical production strategies may decrease the dependence of operation on damaged installation. These adaptation strategies can be used alone or in combination according to the adaptation capabilities of chemical plants. To increase the adjustment strategies, adaptation measures should be taken in advance. By the available adaptation strategies, the loss of chemical plant performance caused by the disruption and the sequential cascading effect may be partially or fully recovered, as follows:

$$f_3 = f_2 + f_{ad} \tag{7.6}$$

f_3 represents the chemical plant performance after adaptation; f_{ad} denotes the increased performance induced by adaptation strategies.

7.3.4 Restoration Modeling

The degradation of performance may be partly, fully, or overly recovered by restoring the damaged chemical plant. In this chapter, all the damaged installations are considered to be reconstructed. In this stage, the key indicator to character the restoration capability is the time to full recovery (TTR) since the reconstruction of installations is a complex and time-consuming process. The restoration capability increases with decreasing the construction time. For instance, the construction of a hazardous material storage tank includes many steps: installing the tank bottom, installing the hydraulic jacking system, installing the tank roof, assembling and lifting the first ring (top) of the tank wall, assembling and lifting the second ring of the tank wall, installing the accessories. Each step of the construction needs several days. The total construction time depends on many factors, such as the construction method, the number of people, and resources invested in the construction. Besides, the restoration sequence may also influence the TTR if multiple installations are damaged. To shorten the restoration time, a company can invest in restoration capability such as formulating restoration plans, establishing a maintenance department, and reach cooperation with reliable construction companies. After restoration, the chemical plant performance can be expressed as follows:

$$f_4 = f_3 + f_{re} \tag{7.7}$$

f_4 represents the chemical plant performance after adaptation; f_{re} denotes the increased performance induced by restoration. In this chapter, f_4 is assumed to equal to f_1.

7.4 Simulation Algorithm

Based on the quantification framework illustrated in Sect. 7.3, this section develops an algorithm to obtain the resilience of chemical plants exposed to disruptions. Due to the uncertainties in the disruption stage and the escalation stage, dynamic Monte Carlo is employed to generate resilience evolution scenarios. Figure 7.3 shows the flow diagram of the dynamic-stochastic algorithm. Firstly, initial parameters need to be inputted in the program, such as the number of iterations N, the disruption time $t_1 = 0$, the initial iteration $n = 1$. Besides, the needed information and data related to the chemical plant are also required, such as the performance of the chemical plant, the installations, and the layout of the chemical plant. Given a disruption, vulnerability analysis will be conducted by using the method illustrated in Sect. 7.3.1. Based on the vulnerability analysis, we can obtain the failure probability of installations exposed to the disruption. According to the failure probabilities, a set of random data (between 0 and 1) is sampled to determine the damaged installations. Each installation is assigned a random value. If the random value is less than the failure

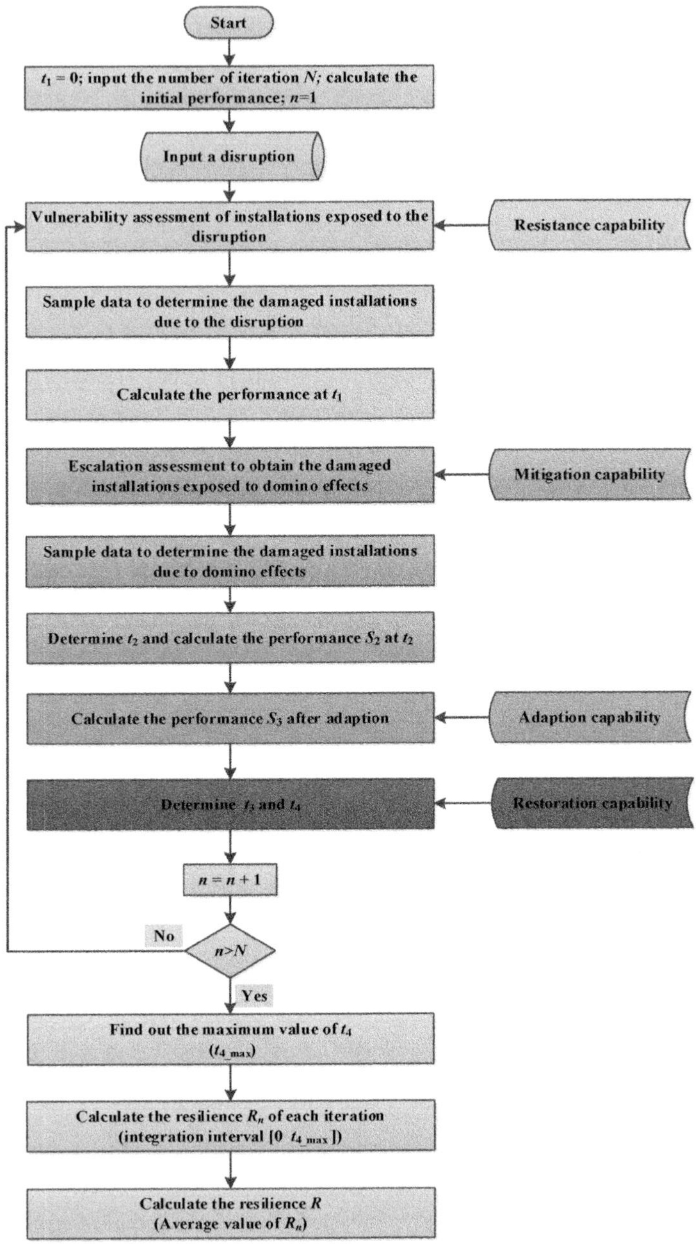

Fig. 7.3 Flow diagram of the algorithm for obtaining chemical plant resilience (Chen et al. [1])

probability, the installation is considered damaged; otherwise, the installation is not damaged. The chemical plant performance should then be updated by subtracting the performance delegation caused by the damaged installations. Next, the possible escalation is assessed using the escalation assessment method illustrated in chapter 2. By the escalation analysis, the failure probability of installations caused by domino effects and the time at the end of the escalation stage can be obtained. According to the results of the escalation analysis, the chemical plant performance at the end of the escalation stage can be calculated. Thirdly, possible adaptation strategies will be determined, and the improved performance induced by the implementation of adaptation measures needs to be determined. Following the adaptation strategy, the restoration strategy should be determined to obtain the start time (t_3) and end time (t_4) of restoration. In that case, a resilience evolution scenario represented by a performance curve (t_1–t_4) can be obtained. The above steps will be repeated until n exceeds N. When the loop ends, we can obtain a total of N resilience evolution scenarios. The maximum value of t_4 (t_{max}) of all the evolution scenarios will be regarded as the upper limit of the integral interval for calculating the resilience value of each scenario. According to the integral interval $[0, t_{max}]$ and the resilience metrics illustrated in Sect. 7.2.3, the resilience value of each scenario can be obtained. Finally, the resilience can be obtained according to Eq. (7.2), considering the dynamic resilience evolution process and uncertainties in the disruption and escalation stages.

7.5 Case Study

In this section, the procedures of the developed resilience approach are illustrated by a case of a chemical storage plant. There are two fuels (gasoline and diesel) in 14 storage tanks (numbering T1–T14). Figure 7.4 shows the layout of the chemical storage plant and Table 7.1 lists the characteristics of the 14 storage tanks. The total storage volume in the initial stage (initial) is 30500 m^3. The flow rate of gasoline and diesel is 1088.6 m^3/d (3×10^8 kg/y) and 658.6 m^3/d (2×10^8 kg/y), respectively. The inventory turnover ratio of the storage farm is 20.9. Assuming that the average value of *TER* (μ) is 15 min and the variance (σ) is 5 min.

Assume that a disruption of an intentional explosion occurs in the chemical storage plant, represented by a red asterisk in Fig. 7.4. The attack is induced by a suitcase bomb with an improvised explosive device (IED) and the explosion is assumed to be equivalent to 23 kg TNT [71]. Once the attack results in tank damage, a suspended time (t_2–t_3) of 30 days is assumed for incident investigation, preparation for restoration. In the restoration stage, the damaged tanks are sequentially rebuilt according to the tank volume (descending order). According to the illustrated information and data. steps to assess the resilience of the chemical storage plant will be illustrated below.

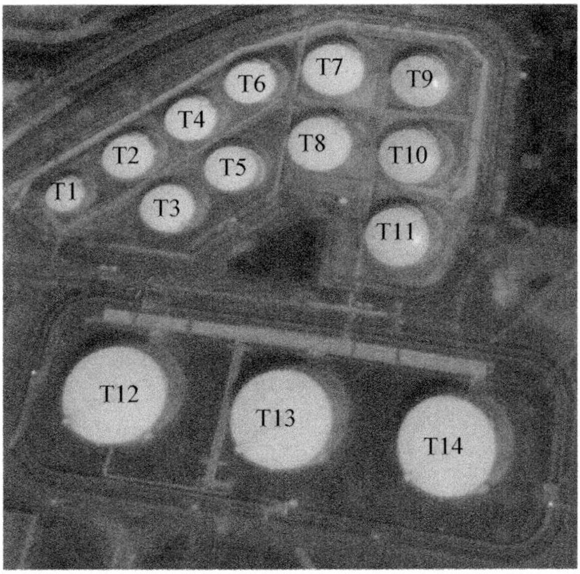

Fig. 7.4 Layout of the oil tank farm (Chen et al. [1])

Table 7.1 Characteristics of storage tanks

Tank	Type	Dimensions (m)	Volume (m^3)	Material	Density (kg/m^3)	Restoration time (day)
T1	Atmospheric, fixed-roof	8.9 × 8.0	500	Diesel	832	15
T2–T5	Atmospheric, floating-roof	11.5 × 12.0	1000	Gasoline	755	30
T6	Atmospheric, fixed-roof	11.5 × 9.6	1000	Diesel	832	30
T7–T11	Atmospheric, fixed-roof	15.7 × 10.4	2000	Diesel	832	60
T12–T14	Atmospheric, floating-roof	21.7 × 16.0	5000	Gasoline	755	150

7.5.1 Resistance Analysis

According to the resilience-based approach, the first step is to analyze the resistance capability to a disruption. The failure probability of tanks exposed to the explosion can be obtained by using the TNT equivalency method and probit models illustrated in Chap. 3. Figure 7.5 shows the damage probability of tanks exposed to the explosion disruption.

As shown in Fig. 7.5, the tanks (T3 and T12) close to the explosion point have much higher damage probabilities than other tanks. In other words, T3 and T12 are more likely to be directly damaged by the explosion while other tanks have a high

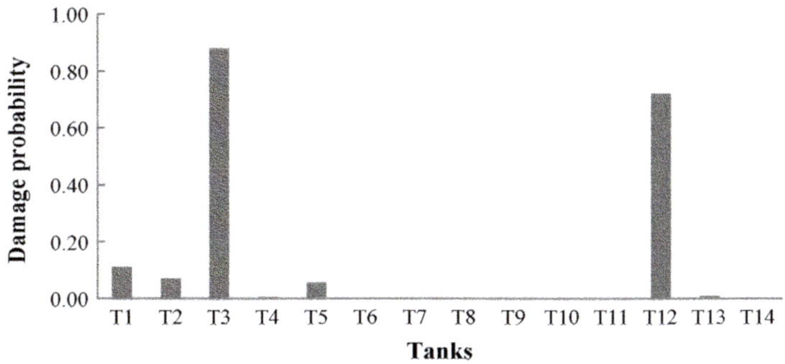

Fig. 7.5 Damage probability of tanks exposed to an IED

probability to survive in the disruption stage. According to the developed algorithm, in a resilience evolution scenario, the damaged tanks can be obtained by sampling 14 random numbers (between 0 and 1) and comparing the random numbers with the damage probabilities, respectively. Given a resilience evolution scenario that only T12 is damaged by the explosion, the total storage capability decreases from 30,500 to 25,500 m^3.

7.5.2 *Mitigation Analysis*

Although only T3 and T12 have a high damage probability (>0.5) in the disruption stage while other tanks may be damaged by domino effects in the escalation stage. The explosion may induce fire on the damaged tanks and the heat radiation generated by the damaged tank to other tanks is obtained by using the ALOHA software [72], as shown in Table 7.2.

According to the potential heat radiation intensities, fire-escalation can be analyzed using the dynamic graph approach developed in Chap. 2. For instance, if only T12 in a resilience evolution scenario is damaged in the disruption stage according to the developed algorithm, then the following fire on T12 may trigger a fire scenario on T13 and subsequently leads to fires on T3 and T14, as shown in Fig. 7.6. T12 is damaged by the explosion, resulting in a fire on T12 at $T = 0$ min. Then, the heat radiation generated by T12 leads to a fire on T13 at $T = 11.5$ min, followed by T3 at $T = 16.5$ min and T14 at $T = 19.9$ min. Due to possible domino effects, the storage performance may further decrease. Given the instance shown in Fig. 7.6, the storage capacity decreases from 25,500 to 14,500 m^3.

Table 7.2 The heat radiation (kW/m^2) from installation *i* to installation *j* (Chen et al. [1])

Tank *i*, *j*	1	2	3	4	5	6	7	8	9	10	11	12	13	14
1	0	11	6	0	0	0	0	0	0	0	0	0	0	0
2	19	0	18	18	11	6	0	0	0	0	0	0	0	0
3	11	18	0	11	18	0	0	0	0	0	0	0	0	0
4	6	18	11	0	18	18	6	7	0	0	0	0	0	0
5	0	11	19	19	0	11	5	13	0	0	0	0	0	0
6	0	0	0	15	9	0	13	13	0	0	0	0	0	0
7	0	0	0	8	8	21	0	18	18	11	0	0	0	0
8	0	5	7	10	18	21	18	0	11	17	11	0	0	0
9	0	0	0	0	0	6	18	11	0	17	5	0	0	0
10	0	0	0	0	6	5.5	11	17	17	0	18	0	0	0
11	0	0	0	0	6	0	0	11	5	18	0	0	0	0
12	9	7	12	5	7	0	0	0	0	0	5	0	17	5
13	0	0	8	0	6.3	0	0	5	0	5	10	17	0	18
14	0	0	0	0	0	0	0	0	0	5	10	5	18	0

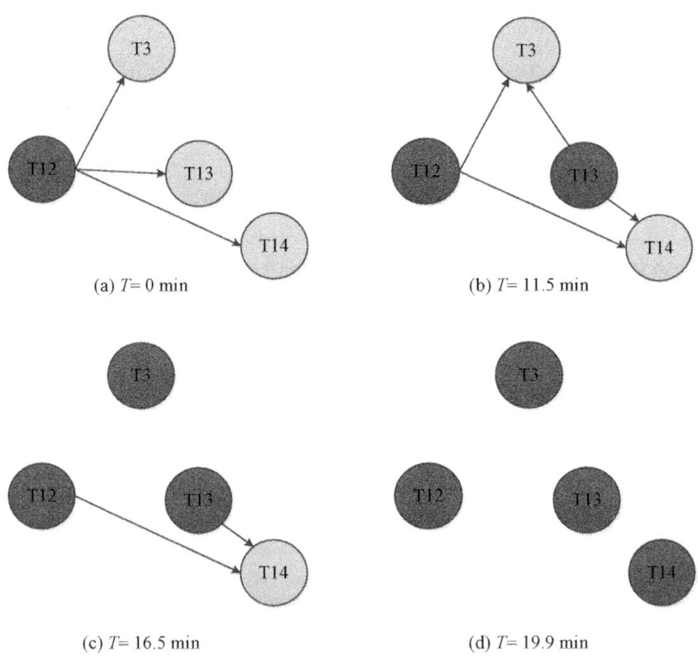

Fig. 7.6 A domino effect caused by the damage of T12 (Chen et al. [1])

7.5.3 Adaptation Analysis

Adaption measures such as utilizing reserve tanks and speeding inventory turnover ratio can compensate for the performance loss caused by a disruption before the storage plant is restored. However, the adaption capability is limited by the storage equipment (reserve tanks), loading, and unloading facilities. The performance loss may not be fully restored by using adaptation strategies. In this case study, the inventory turnover is assumed to be increased by 20% after the escalation stage. Following the instance in Sect. 7.5.2, the storage capacity increases from 14,500 to 17,400 m^3 by speeding the inventory turnover ratio.

7.5.4 Restoration Analysis

Restoration is the final stage of a resilience evolution scenario. According to the restoration time of different tanks shown in Table 7.1, we can obtain the entire performance curve of a resilience evolution scenario, as shown in Fig. 7.7. The storage plant's storage performance decreases from 30,500 to 25,500 m^3 due to the damage of T12 exposed to the explosion at $T = 0$ min. In the escalation stage, the storage performance further decreases from 25,500 to 14,500 m^3 since T13, T3, and T14 are sequentially damaged by the domino effect. At the end of the escalation stage, an adaptation strategy is used, resulting in an increase of storage performance from 14,500 to 17,400 m^3. Finally, T12, T13, T14, and T3 are sequentially restored, leading to a recovery of storage performance.

Figure 7.7 only shows one possible resilience scenario. Applying the developed algorithm in this chapter, we can obtain many evolution scenarios due to uncertainties in resilience. Then, according to Eq. (7.2), we can obtain the chemical storage plant's resilience value of 0.85.

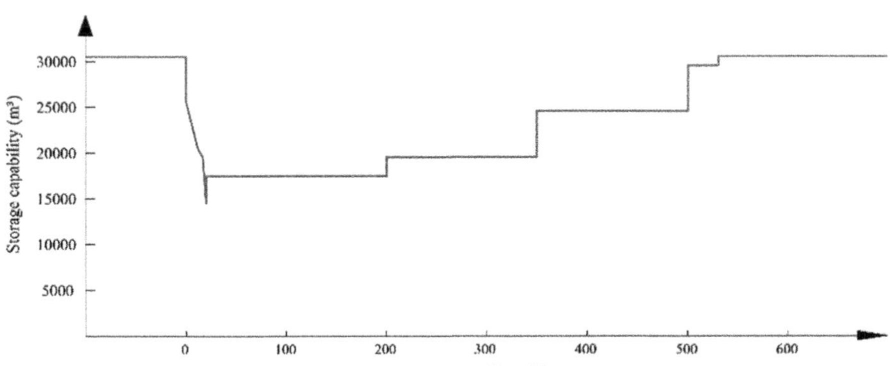

Fig. 7.7 Performance curve of a resilience evolution scenario (Chen et al. [1])

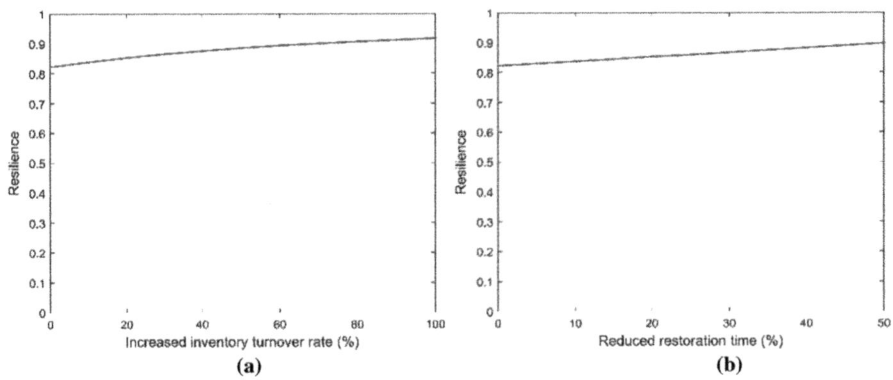

Fig. 7.8 Resilience performance of **a** speeding inventory turnover ratio and **b** reducing restoration time

7.5.5 *Resilience Measure Performance*

According to the resilience-based approach developed in this chapter, the performance of different resilience measures can be quantified to support resilience management. Figure 7.8 shows the performance of (a) speeding inventory turnover ratio and (b) reducing restoration time on the chemical storage plant's resilience. The resilience increases with increasing the inventory turnover ratio and with decreasing restoration time. Besides adaptation restoration measures, safety barriers can also be used to enhancing resilience by preventing and mitigating domino effects. Figure 7.9 shows the resilience performance of a safety barrier (emergency response). The resilience increases with the reduction of the mean value of the time needed for emergency response (u).

7.6 Conclusions

This chapter proposes a dynamic stochastic approach to quantify the resilience of chemical plants, considering possible escalation effects, adaptation strategies, and the recovery of damaged chemical plants. In this approach, a dynamic stochastic resilience process (called a resilience evolution scenario) is divided into four stages: disruption, escalation, adaption, and restoration stages. The uncertainties in the vulnerability of installations exposed to disruptions, major accident escalations, and emergency response are considered. A dynamic Monte Carlo method is used to generate possible resilience evolution scenarios and obtain chemical plant resilience. Compared with traditional safety and security management methods, the resilience-based method highlights the roles of adaptation capability and restoration capability in tackling unpredictable and unpreventable disruptions. According to the developed approach, the uncertainties in the resilience process cannot be ignored; domino

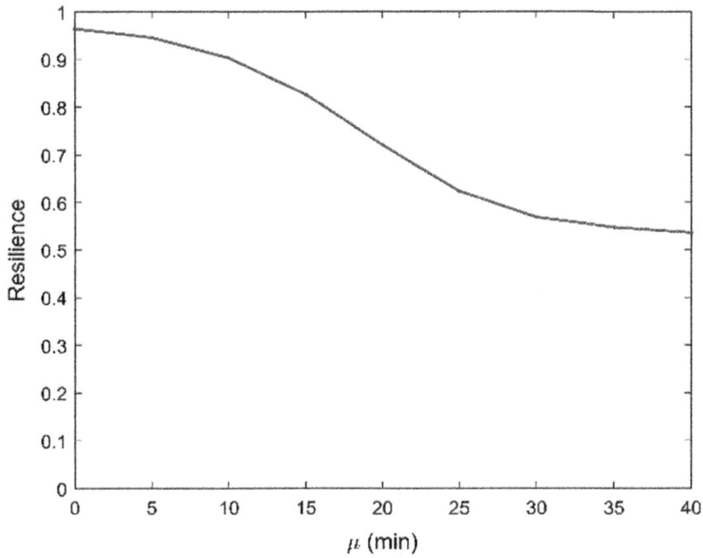

Fig. 7.9 Resilience performance of enhancing emergency response capability

effects play an essential role in chemical plant resilience; All the resilience measures, such as safety barriers used in the escalation stage, speeding inventory turnover in the adaptation stage, and shortening the restoration time in the restoration stage is effective for dealing with domino effects.

References

1. Chen C, Yang M, Reniers G (2021) A dynamic stochastic methodology for quantifying HAZMAT storage resilience. Reliab Eng Syst Saf
2. Chen C (2021) A dynamic and integrated approach for modeling and managing domino-effects. Delft University of Technology
3. Chen C, Reniers G, Khakzad N (2020) A thorough classification and discussion of approaches for modeling and managing domino effects in the process industries. Saf Sci 125
4. Fuentes-Bargues JL, González-Cruz M, González-Gaya C, Baixauli-Pérez M (2017) Risk analysis of a fuel storage terminal using HAZOP and FTA. Int J Environ Res Public Health 14(7):705
5. Yang Y, Chen G, Chen P (2018) The probability prediction method of domino effect triggered by lightning in chemical tank farm. Process Saf Environ Prot 116:106–114
6. Tong Q, Yang M, Zinetullina A (2020) A dynamic Bayesian network-based approach to resilience assessment of engineered systems. J Loss Prevent Process Ind 65
7. Pasman H, Kottawar K, Jain P (2020) Resilience of process plant: what, why, and how resilience can improve safety and sustainability. Sustainability 12(15)
8. Khan F, Rathnayaka S, Ahmed S (2015) Methods and models in process safety and risk management: past, present and future. Process Saf Environ Prot 98:116–147

9. Reniers GLL, Audenaert A (2014) Preparing for major terrorist attacks against chemical clusters: intelligently planning protection measures w.r.t. domino effects. Process Saf Environ Protect 92(6):583–589
10. Cozzani V, Antonioni G, Landucci G, Tugnoli A, Bonvicini S, Spadoni G (2014) Quantitative assessment of domino and NaTech scenarios in complex industrial areas. J Loss Prev Process Ind 28:10–22
11. CNN (2019) Texas district attorney files charges against chemical plant for its 4-day fire. https://edition.cnn.com/2019/04/29/us/deer-park-itc-plant-fire-charges/index.html
12. Cozzani V, Campedel M, Renni E, Krausmann E (2010) Industrial accidents triggered by flood events: analysis of past accidents. J Hazard Mater 175(1–3):501–509
13. Khakzad N, Van Gelder P (2018) Vulnerability of industrial plants to flood-induced natechs: a Bayesian network approach. Reliab Eng Syst Saf 169:403–411
14. Qin R, Khakzad N, Zhu J (2020) An overview of the impact of Hurricane Harvey on chemical and process facilities in Texas. Int J Disast Risk Reduct 45
15. Misuri A, Casson Moreno V, Quddus N, Cozzani V (2019) Lessons learnt from the impact of hurricane Harvey on the chemical and process industry. Reliab Eng Syst Saf 190
16. Chen C, Reniers G, Khakzad N (2020) Cost-benefit management of intentional domino effects in chemical industrial areas. Process Saf Environ Prot 134:392–405
17. Khan FI, Amyotte PR (2003) How to make inherent safety practice a reality. Canad J Chem Eng 81(1):2–16
18. Cozzani V, Tugnoli A, Salzano E (2007) Prevention of domino effect: from active and passive strategies to inherently safer design. J Hazard Mater 139(2):209–219
19. Landucci G, Tugnoli A, Cozzani V (2008) Inherent safety key performance indicators for hydrogen storage systems. J Hazard Mater 159(2–3):554–566
20. Eini S, Abdolhamidzadeh B, Reniers G, Rashtchian D (2015) Optimization procedure to select an inherently safer design scheme. Process Saf Environ Prot 93:89–98
21. Tanabe M, Miyake A (2012) Approach enhancing inherent safety application in onshore LNG plant design. J Loss Prev Process Ind 25(5):809–819
22. Khan FI, Husain T, Abbasi SA (2001) Safety weighted hazard index (SWeHI): a new, user-friendly tool for swift yet comprehensive hazard identification and safety evaluation in chemical process industries. Process Saf Environ Prot 79(2):65–80
23. Cameron I, Mannan S, Németh E, Park S, Pasman H, Rogers W, Seligmann B (2017) Process hazard analysis, hazard identification and scenario definition: are the conventional tools sufficient, or should and can we do much better? Process Saf Environ Prot 110:53–70
24. Guo X, Ji J, Khan F, Ding L (2020) Fuzzy Bayesian network based on an improved similarity aggregation method for risk assessment of storage tank accident. Process Saf Environ Prot 144:242–252
25. Chen C, Khakzad N, Reniers G (2020) Dynamic vulnerability assessment of process plants with respect to vapor cloud explosions. Reliab Eng Syst Saf 200
26. Villa V, Paltrinieri N, Khan F, Cozzani V (2016) Towards dynamic risk analysis: a review of the risk assessment approach and its limitations in the chemical process industry. Saf Sci 89:77–93
27. Cozzani V, Gubinelli G, Antonioni G, Spadoni G, Zanelli S (2005) The assessment of risk caused by domino effect in quantitative area risk analysis. J Hazard Mater 127(1–3):14–30
28. Antonioni G, Landucci G, Necci A, Gheorghiu D, Cozzani V (2015) Quantitative assessment of risk due to NaTech scenarios caused by floods. Reliab Eng Syst Saf 142:334–345
29. Yang Y, Chen G, Reniers G (2019) Vulnerability assessment of atmospheric storage tanks to floods based on logistic regression. Reliab Eng Syst Saf 106721
30. Zhou J, Reniers G, Zhang L (2017) A weighted fuzzy Petri-net based approach for security risk assessment in the chemical industry. Chem Eng Sci 174:136–145
31. Reniers G, Van Lerberghe P, Van Gulijk C (2015) Security risk assessment and protection in the chemical and process industry. Process Saf Prog 34(1):72–83
32. Khakzad N, Martinez IS, Kwon HM, Stewart C, Perera R, Reniers G (2018) Security risk assessment and management in chemical plants: challenges and new trends. Process Saf Prog 37(2):211–220

33. Matteini A, Argenti F, Salzano E, Cozzani V (2018) A comparative analysis of security risk assessment methodologies for the chemical industry. Reliab Eng Syst Saf
34. Gnoni MG, Lettera G, Bragatto PA (2009) A decision model for preventing knock-on risk inside industrial plant. In: Safety, reliability and risk analysis: theory, methods and applications, vols 1–4. CRC Press-Taylor & Francis Group, Boca Raton
35. Reniers G, Dullaert W, Karel S (2009) Domino effects within a chemical cluster: a game-theoretical modeling approach by using Nash-equilibrium. J Hazard Mater 167(1–3):289–293
36. Reniers G, Pavlova Y (2013) Safety management in chemical industrial clusters: the state of the art
37. Reniers G, Dullaert W, Audenaert A, Ale BJ, Soudan K (2008) Managing domino effect-related security of industrial areas. J Loss Prev Process Ind 21(3):336–343
38. Zhang L, Reniers G (2016) A game-theoretical model to improve process plant protection from terrorist attacks. Risk Anal 36(12):2285–2297
39. Khakzad N, Reniers G (2019) Low-capacity utilization of process plants: a cost-robust approach to tackle man-made domino effects. Reliab Eng Syst Saf 191
40. Hosseinnia B, Khakzad N, Reniers G (2018) Multi-plant emergency response for tackling major accidents in chemical industrial areas. Saf Sci 102:275–289
41. Zhou J (2013) Petri net modeling for the emergency response to chemical accidents. J Loss Prev Process Ind 26(4):766–770
42. Du Y, Sun J, Duan Q, Qi K, Xiao H, Liew KM (2020) Optimal assignments of allocating and scheduling emergency resources to accidents in chemical industrial parks. J Loss Prevent Process Ind 65
43. Chen C, Reniers G, Khakzad N (2019) Integrating safety and security resources to protect chemical industrial parks from man-made domino effects: a dynamic graph approach. Reliab Eng Syst Saf 191
44. Reniers G (2009) Man-made domino effect disasters in the chemical industry: the need for integrating safety and security in chemical clusters. Disaster Adv 2(2):3–5
45. Song G, Khan F, Yang M (2018) Integrated risk management of hazardous processing facilities. Process Saf Progress
46. Reniers G, Cozzani V (2013) Domino effects in the process industries, modeling, prevention and managing. Elsevier, Amsterdam, The Netherlands
47. Hosseini S, Barker K, Ramirez-Marquez JE (2016) A review of definitions and measures of system resilience. Reliab Eng Syst Saf 145:47–61
48. Doorn N, Gardoni P, Murphy C (2018) A multidisciplinary definition and evaluation of resilience: the role of social justice in defining resilience. Sustain Resil Infrastruct 4(3):112–123
49. Holling CS (1973) Resilience and stability of ecological systems. Annu Rev Ecol Syst 4(1):1–23
50. National Infrastructure Advisory Council (2009) Critical infrastructure resilience: final report and recommendations. National Infrastructure Advisory Council
51. Hosseini S, Barker K (2016) Modeling infrastructure resilience using Bayesian networks: a case study of inland waterway ports. Comput Ind Eng 93:252–266
52. Yarveisy R, Gao C, Khan F (2020) A simple yet robust resilience assessment metrics. Reliab Eng Syst Saf 197
53. Cincotta S, Khakzad N, Cozzani V, Reniers G (2019) Resilience-based optimal firefighting to prevent domino effects in process plants. J Loss Prev Process Ind 58:82–89
54. Zinetullina A, Yang M, Khakzad N, Golman B, Li X (2021) Quantitative resilience assessment of chemical process systems using functional resonance analysis method and dynamic Bayesian network. Reliab Eng Syst Saf 205
55. Provan DJ, Woods DD, Dekker SWA, Rae AJ (2020) Safety II professionals: how resilience engineering can transform safety practice. Reliab Eng Syst Saf 195
56. Hollnagel E, Woods DD, Leveson N (2006) Resilience engineering: concepts and precepts. Ashgate Publishing, Ltd.
57. Dinh LTT, Pasman H, Gao X, Mannan MS (2012) Resilience engineering of industrial processes: principles and contributing factors. J Loss Prev Process Ind 25(2):233–241

58. Jain P, Mentzer R, Mannan MS (2018) Resilience metrics for improved process-risk decision making: survey, analysis and application. Saf Sci 108:13–28
59. Jain P, Rogers WJ, Pasman HJ, Keim KK, Mannan MS (2018) A Resilience-based Integrated Process Systems Hazard Analysis (RIPSHA) approach: part I plant system layer. Process Saf Environ Prot 116:92–105
60. Alrabghi A (2020) Modelling condition-based maintenance for nonidentical two-component systems considering four types of dependencies. J Qual Main Eng ahead-of-print (ahead-of-print).
61. Azadeh A, Salehi V, Arvan M, Dolatkhah M (2014) Assessment of resilience engineering factors in high-risk environments by fuzzy cognitive maps: a petrochemical plant. Saf Sci 68:99–107
62. Abimbola M, Khan F (2019) Resilience modeling of engineering systems using dynamic object-oriented Bayesian network approach. Comput Ind Eng 130:108–118
63. Reniers GLL, Sörensen K, Khan F, Amyotte P (2014) Resilience of chemical industrial areas through attenuation-based security. Reliab Eng Syst Saf 131:94–101
64. Bruneau M, Chang SE, Eguchi RT, Lee GC, O'Rourke TD, Reinhorn AM, Shinozuka M, Tierney K, Wallace WA, Von Winterfeldt D (2003) A framework to quantitatively assess and enhance the seismic resilience of communities. Earthq Spectra 19(4):733–752
65. Pant R, Barker K, Ramirez-Marquez JE, Rocco CM (2014) Stochastic measures of resilience and their application to container terminals. Comput Ind Eng 70:183–194
66. Henry D, Emmanuel Ramirez-Marquez J (2012) Generic metrics and quantitative approaches for system resilience as a function of time. Reliab Eng Syst Saf 99:114–122
67. Necci A, Antonioni G, Bonvicini S, Cozzani V (2016) Quantitative assessment of risk due to major accidents triggered by lightning. Reliab Eng Syst Saf 154:60–72
68. Landucci G, Argenti F, Tugnoli A, Cozzani V (2015) Quantitative assessment of safety barrier performance in the prevention of domino scenarios triggered by fire. Reliab Eng Syst Saf 143:30–43
69. Khakzad N, Landucci G, Reniers G (2017) Application of dynamic Bayesian network to performance assessment of fire protection systems during domino effects. Reliab Eng Syst Saf 167:232–247
70. Johansson J, Hassel H, Zio E (2013) Reliability and vulnerability analyses of critical infrastructures: comparing two approaches in the context of power systems. Reliab Eng Syst Saf 120:27–38
71. Hosseinnia B, Khakzad N, Reniers G (2018) An emergency response decision matrix against terrorist attacks with improvised device in chemical clusters. Int J Saf Secur Eng 8(2):187–199
72. ALOHA user manual (2016) US Environmental Protection Agency, National Oceanic and Atmospheric Administration, ALOHA, Version 5.4.7. https://www.epa.gov/cameo/aloha-software. Accessed 5 May 2019

Chapter 8
Conclusions and Future Research

This book introduces three domino effect models that can be used for risk assessment of both intentional domino effects and unintentional domino effects. Based on these risk assessment models, three chapters on domino effect management are provided to prevent and mitigate domino effects triggered by unintentional events or intentional attacks.

When a hazardous scenario (toxic release, fire, and explosion) occurs in a chemical industrial area, many hazardous installations are mutually linked via escalation vectors (e.g., heat radiation, overpressure), forming a system [1]. The spatial–temporal evolution of hazardous scenarios within the system may lead to domino effects. In light of the characteristics, a dynamic tool is better to model the temporal evolution and a graph/network-based approach is suitable to model the spatial escalation. As a result, a domino evolution graph (DEG) model based on dynamic graphs is proposed in Chap. 2 to model the spatial–temporal evolution of fire-induced domino effects. In Chap. 3, a dynamic event tree is used to model the dynamic evolution of vapor cloud explosion. Besides, Monte Carlo method is integrated into the dynamic graph model (called "Dynamic Graph Monte Carlo" (DGMC)) in Chap. 4 to tackle the evolution uncertainties in domino effects. In the DGMC model, hazardous installations, humans, and ignition sources are modeled as graph nodes and the physical effects between different nodes are modeled as graph edges. The Monte Carlo method is employed to automatically update graphs and deal with uncertainties, obtaining numerical evolution results.

In terms of domino effect management, not only safety barriers but also security measures should be considered due to possible catastrophic consequences caused by intentional attacks. In decision-making on prevention and mitigation of domino effects, economic issues need to be considered because chemical companies usually face budget limitations and pursue more profit. As a result, safety and security resources are integrated into domino effect management in Chap. 5, and a cost–benefit analysis is conducted in Chap. 6 to obtain the optimal protection strategy. Besides safety and security measures, adaptation and restoration should also be considered

C. Chen et al., *Integrating Safety and Security Management to Protect Chemical Industrial Areas from Domino Effects*, Springer Series in Reliability Engineering,
https://doi.org/10.1007/978-3-030-88911-1_8

to deal with unpredicted and unpreventable domino effects. Therefore, a resilience-based approach is developed in Chap. 7 in which the roles of resistance, mitigation, adaptation, and restoration in domino effect management are quantified.

8.1 Main Conclusions

(1) Conclusions on the state-of-the-art of domino effect research

Since the 1990s, increasing attention has been paid to domino effects in the process industry. In the past three decades, various methods have been developed to model and manage domino effects in the process industry. The modeling methods were divided into three categories: analytical approaches, graphical approaches, and simulation approaches. Some analytical methods-based software was developed in early research of domino effects to quantify the likelihood of domino effects. Graphical approaches, such as Bayesian network and Petri-net obtain increasing attention in recent years and can be used to map higher-order escalation of domino effects and thus estimate the probability of domino effects and the vulnerability of installations. Simulation approaches based on the Monte Carlo method can simplify probability calculation and may be used to deal with complex escalations while requiring longer calculation time. Although current methods have contributed enormously to modeling domino effects, many challenges still exist. The problems include modeling the spatial–temporal evolution of domino effects involving higher-order escalations, modeling the VCE-induced domino effects, and modeling the evolution of multi-hazardous scenarios in one domino effect. In terms of the prevention and mitigation of domino effects, various management strategies were proposed: inherent safety, management of safety barriers, emergency response, cooperative prevention, and security strategies. Safety managers may select one protection strategy or a combination of multiple strategies. These strategies with different performances and costs may be used in different stages of the entire operating life. Thus, both the protection costs and financial implications related to potential avoided losses should be considered since protection resources are always limited and essential for the company's profitability in the long term. Besides, both safety and security measures should be used to deal with intentional domino effects in which multiple failures of installations are possible. Once domino effects are inevitable, a quick repair or reconstruction may reduce the consequences of domino effects. As a result, the adaptation and restoration of chemical plants should also be considered in the whole to deal with unpredictable or unpreventable domino effects.

(2) Conclusions on risk assessment of fire-induced domino effects

In this book, a domino evolution graph (DEG) model based on dynamic graphs is developed in Chap. 2. In the model, hazardous installations are modeled as graph nodes and the escalation vectors are modeled as graph edges. The graph structure can model possible complex phenomena in spatial evolution, such as synergistic effects and parallel effects. Besides, graph updates can model the time dependencies in temporal evolution such as superimposed effects. Moreover, the model can also overcome the limitation of the probit model in higher-level escalation and rapidly obtain the evolution paths, evolution time, and the failure probabilities of installations. An illustrated case study demonstrates that the synergistic effects and the superimposed effects considered in the DEG play a vital role in domino effect evolution. The domino effect risk may be underestimated if they are ignored in domino effect modeling. The primary scenario involving the failure of multiple installations that are more likely to occur in intentional domino effects can speed up the escalation of domino effects, leading to the prevention of domino effects more difficult and more severe consequences. Since the evolution process and the damage probability of installations can be rapidly obtained using the dynamic graph approach, the developed model can be applied to realistic chemical clusters with a large number of installations, significantly supporting the decision-making on the allocation of safety and security resources.

(3) **Conclusions on risk assessment of VCE-induced domino effects**

Chapter 3 establishes a dynamic VCE evolution assessment (DVEA) model based on a dynamic event tree. The DVEA model integrates the dispersion process of vapor cloud and ignition uncertainty into a stochastic simulation engine (a dynamic event tree) to assess the vapor cloud explosion risk in chemical industrial areas and obtain the damage probabilities of installations exposed to VCEs and the likelihood of domino effects. Both the time dependencies in vapor cloud dispersion and the uncertainty of delayed ignitions are addressed in the DVEA model. Applying the DVEA model in a real case shows that both the time dependencies in vapor cloud dispersion and the uncertainty of delayed ignition are crucial for reflecting the characteristics of possible large VCEs and avoiding underestimating consequences. The vulnerability of installations to VCEs depends on the congestion of the plant layout and delayed ignition time (DIT). A long-delayed explosion may result in multiple-failure of installations, resulting in catastrophic disasters. The influence factors of DIT include the distance between the release position and the ignition sources, the type of ignition sources, and the ignition control measures in place. Ignition control measures in a chemical plant can decrease the ignition probability of single sources while may lead to a larger VCE and more severe consequences if the vapor cloud disperses outside the plant in which ignition sources are not fully eliminated. As a result, ignition control may be regarded as a delay measure but not as a preventive measure. Combining ignition control measures with emergency response actions may be an effective way to prevent VCEs since ignition control might provide enough time for emergency response actions to prevent VCEs.

(4) **Conclusions on multiple accident scenarios in a domino effect**

Based on the research of the DEG model of fire-induced domino effects in Chap. 2 and the DVEA model of VCE-induced domino effects in Chap. 3, a dynamic model called "Dynamic Graph Monte Carlo" (DGMC) is developed to model the evolution of multi-hazardous scenarios and assess the vulnerability of humans and installations exposed to such hazards. Since the DGMC model integrates dynamic graphs and Monte Carlo method, it has the advantages of both methods: graphs can provide a structure for model spatial evolution, graph update can model temporal evolution, and the random number generator in Monte Carlo simulation can deal with uncertainties in domino effect evolution. Therefore, the DGMC model can effectively model multiple hazardous scenarios that may simultaneously or sequentially occur in one domino effect. Neglecting any hazardous scenarios may underestimate the consequences of domino effects, resulting in an unreasonable allocation of safety barriers and personal protection equipment (PPE). The study results show that humans in different locations may be threatened by different hazards, thus different protection measures may be formulated for different people. A long-delayed ignition can damage multiple installations and acute toxicity of people around the release source. As a result, VCE-induced domino effects may result in more severe consequences than fire-induced domino effects. The safety distances based on fire hazards are not sufficient for the prevention of VCE-induced domino effects. People close to the release source are prone to the threat of multi-hazardous scenarios, while the distant deaths are mainly induced by acute toxicity and overpressure.

(5) **Conclusions on integrated safety and security management of domino effects**

Intentional attacks may lead to simultaneous accident scenarios, resulting in synergistic effects and making the prevention of domino effects more difficult than unintentional domino effects. As a result, an integrated safety and security management framework is developed in Chap. 5, considering the performance of both safety measures and security measures. This framework includes six steps: chemical plant description, threat, and hazard identification, the vulnerability of installations subject to hazards and threats, the vulnerability of installations exposed to domino effects, consequence analysis, risk treatment, and risk reduction. These measures are divided into three categories: detection measures, delay measures, and response measures. According to the functions of each protection measure and domino effect models proposed in this book, the performance of a protection strategy on domino effect risk reduction can be obtained, considering fatalities, property loss, environmental impacts, business loss, and reputation or negative publicity. Finally, the domino effect risk can be calculated and evaluated to determine whether the risk is acceptable. If the risk is unacceptable, additional protection measures should be taken until the risk is lower than the pre-defined risk tolerability criteria. According to this framework, decision-making approaches can be developed to obtain an optimal protection strategy.

(6) **Conclusions on cost–benefit management of domino effects**

According to the integrated management framework developed in Chap. 5, a cost–benefit management approach is established to support decision-making on protection measures. The performance of a protection strategy (a combination of different protection measures) is quantified and monetized in the cost–benefit analysis. The expected avoided loss caused by a protection strategy is considered the benefit while the investment related to the protection strategy is regarded as the cost. As a result. the disproportion factor (DF) is employed in the cost–benefit analysis to determine whether a protection strategy is recommended. Besides, an optimization algorithm called "PROTOPT" is developed to achieve the optimal strategy. The study demonstrates that multiple kinds of protection measures should be employed in chemical industrial areas since they follow the law of diminishing returns. The likelihood of threats plays a critical role in a protection strategy's benefits. Therefore the optimal protection strategy varies with different plants and different threats. The protection is profitable only when the threat likelihood is no less than the threat probability at the break-even point. At the break-even point, the protection benefit is equal to the protection cost.

(7) **Conclusions on resilience management of domino effects**

Domino effects may be unpreventable such as the escalation caused by simultaneous attacks or natural disasters. In that case, protection measures may not prevent domino effects. A feasible way to deal with these unpreventable domino effects is to reduce the effects on the operation of companies by adjusting operation strategies and rapidly restoring the damaged installations. Resilience is the ability of a system to resist, mitigate, adapt and recover from disruptions. As a result, enhancing the resilience of a chemical plant can promote to prevent and mitigate domino effects, adapt, and recover from the damaged situation. A resilience-based approach is thus proposed in Chap. 7 to deal with domino effects. In this chapter, a dynamic stochastic model is developed to quantify the resilience of chemical plants. A resilience evolution scenario is modeled as a dynamic process that consists of four stages: disruption, escalation, adaptation, and restoration stages. A simulation algorithm is developed to generate possible resilience evolution scenarios for obtaining chemical plant resilience. Besides safety and security measures, the developed resilience approach highlights the roles of adaptation and restoration in dealing with domino effects, which is more suitable for tackling unpredictable and unpreventable disruptions. Chemical plant resilience depends on resistance capability, mitigation capability, adaption capability, and restoration capability. Improving any of these capabilities can contribute to the prevention and mitigation of domino effects. Various resilience measures such as safety barriers in the escalation stage, speeding inventory turnover in the adaptation stage, and shortening restoration time in the restoration stage are effective for developing a more resilient chemical plant and thus reducing the likelihood and consequences of domino effects.

8.2 Recommendations for Future Research

(1) Recommendations on probit models

Probit models are used in this study for the risk assessment of domino effects and significantly impact the reasonable risk assessment results. The probit models for assessing fire-induced domino effects depend on the time to failure (TTF) of vessels exposed to fire. The common-used calculation method for TTF is developed for small vessels (atmospheric vessels: 25–17,500 m^3; pressurized vessels: 5–250 m^3). However, using large storage vessels for hazardous materials is a development trend in the process industry [2]. Besides vessel types, volumes considered in probit models, other vessel parameters such as wall thickness and wall material may also impact the vessel vulnerability. To extend the application of probit models and this study, vulnerability experiments may be used to improve the probit models. But experiments of large vessels may be expensive and dangerous, numerical simulations may be conducted using advanced consequence simulation software, such as ANSYS, FLUENT, FLACS, FDS, etc.

(2) Recommendation on uncertainty modeling in domino effects

This study develops graphical-based models for modeling the spatial–temporal evolution of domino effects, addressing the time-dependencies, ignition uncertainty, and possible multiple hazardous scenarios in domino effects. However, accurately modeling domino effects is still challenging due to the uncertainty involved in the evolution of domino effects. The uncertainty can be divided into two parts, the intensity uncertainty of hazardous scenarios and the uncertainty of propagation. The former refers to heat radiation intensity, overpressure value, and the number, weight, and velocity of fragments. The latter involves the failure likelihood of installations subject to hazardous scenarios, failure types, and the subsequent scenarios. These uncertainties may be tackled in future research to support domino effect risk assessment and management.

(3) Recommendation on modeling VCE-induced domino effects

In this study, vapor cloud dispersion is considered in VCE-induced domino effects. However, the vapor cloud dispersion model based on empirical formulas may not extend to model all possible release scenarios. Besides, the empirical model neglects VCE dilution with distance, the influences of wind velocity, and the effects of obstacles on dispersion. Thus, dynamic CFD methods may be integrated into domino effect risk assessment to obtain more accurate results in future studies. With the rapid improvement of computational resources, applying dynamic CFD methods in risk assessment may become easier and acceptable for researchers and practitioners in the future.

(4) Recommendation on domino effect management

In Chap. 6, a management approach is established to prevent and mitigate domino effects in chemical plants. However, there may be multiple chemical plants belonging to different companies in a chemical cluster. In terms of the cross-plant areas, safety and security resources allocated in one chemical plant has a benefit for nearby plants due to the mitigation of possible external domino effects while it may also relatively increase the security risk of nearby plants because of the change of attractiveness for possible common adversaries. To get the optimal strategy in a chemical plant, the protection strategies of other plants should be considered. Hence, the cost–benefit management may be extended to support decision-making on domino effects in chemical luster. Besides, the management approach proposed in this study neglects inherent safety design. Future research may consider inherent safety measures in a protection strategy to develop a life cycle management tool.

(5) Recommendations on economic aspects of safety and security

In this study, cost–benefit analysis is used to support decision-making on protection strategies. The reliability of the optimal protection obtained by economic approaches depends on the monetization of costs. However, economic data is difficult to collect and a database for economic values of accident costs and the costs of safety measures may be developed in the future. Besides, some costs are difficult to be monetized or unethical to be monetized, such as the value of human life, reputational costs, and psychological costs. Therefore, other economic approaches such as cost-effectiveness analysis may be used to reduce the monetization work. Moreover, multiple-criteria decision (MCD) may be developed to deal with these costs and multi-objective optimization may be used to obtain the optimal protection strategy.

(6) Recommendations on resilience-based approach

Chapter 7 develops a resilience-based approach for tackling domino effects, considering safety measures, security measures, adaptation measures, and restoration measures. Besides these measures, more design and operation options may be identified and quantified in the future to improve chemical plant resilience. Furthermore, the costs of different resilience measures are not considered and the benefits of resilience are not monetized. To support decision-making on resilience investment, resilience management approaches may be developed by combining the resilience quantification method developed in this study with economic tools such as cost–benefit analysis and cost-effectiveness analysis. Furthermore, the resilience quantification method developed for chemical plants may be applied to other interdependent infrastructure systems such as water supply systems and energy transportation systems. In recent decades, domino effect accidents in the process industry have raised an increasing concern in scientific and technical domains. Domino effects such as the Buncefield accident in 2005 can lead to a chain of accidents and result in more severe consequences than the primary event. The primary event can be an accident, a natural disaster, or an intentional attack. To prevent and mitigate domino effects, growing research on risk assessment and management of domino effects has been conducted in recent decades such as quantitative risk assessment of domino effects and safety barrier management for preventing domino effects. However, little attention has been paid to domino effects triggered by intentional events (intentional domino effects) such as terrorist attacks. This book thus aims to introduce advanced approaches that can model and manage both intentional and unintentional domino effects, developing a safer, securer, and more resilient process plant.

References

1. Chen C (2021) A dynamic and integrated approach for modeling and managing domino-effects. Delft University of Technology
2. Yang Y, Kim J, Seo H, Lee K, Yoon I (2006) Development of the world's largest above-ground full containment LNG storage tank. In: 23rd World gas conference, Amsterdam, pp 1–14

Printed by Printforce, the Netherlands